ISBN: 978-1-300-73145-0

TEXTBOOKS ON UNIFICATION SCIENCE

THE
UNIFICATION SCIENCES:
MATHEMATICS,
PHYSICS AND CHEMISTRY

by
Dr. Richard L. Lewis

UNIFICATION THOUGHT INSTITUTE
KOREA • JAPAN • USA

TEXTBOOKS ON UNIFICATION SCIENCE

Publisher:

Dr. Jin Sung-Bae
President, UTI-Korea

Editors:

Dr. Jin Sung-Bae,
Rev. Hideo Oyamada,
Akifumi Otani,
Dr. Jung Chang Choi,
Dr. Thomas Ward

Coordinator:

Mr Takeshi Furuta

Editorial Consultants:

Mr Jong-Sam Lee,
Dr. Claude Perrottet,
Dr. Richard Lewis

BOOKS IN THE SERIES

Pursuing the Unity of Sciences
Dr. Sung-Bae Jin

Unification Physics
Dr. Ching-Ching Chang

The Unification Sciences: **Mathematics, Physics and Chemistry**
Dr. Richard Lewis

Unification Medical Science
Dr. Shigehiro Suzuki

Unification Perspectives on Peace and Conflict Transformation
Dr. Thomas Ward and Dr. Claude Perrottet

Unification Ethics of True Love
Akifumi Otani

FOREWORD

I would like to propose a few characteristics of the philosophy that can lead world civilization. First, the philosophy to lead world civilization must have a firm moral framework. The progress of society and culture cannot be measured in terms of technological advances or material superiority. Rather, the maturity of a society and culture should be measured by its standard of goodness. Second, the philosophy to lead world civilization must be able to embrace both the civilizations of the East and West and generate a new civilization through their integration. It should be able to harmonize the technical culture based on the analytical and logical thinking of the West and an ethical culture based on the intuitive thinking of the East. Third, the philosophy to lead world civilization must discover God, lost in modern civilization, and adopt absolute values centered on God as its foundation.

Out of an excessive emphasis on human reason, Western civilization has dethroned God replacing Him with a humanist philosophy, even promulgating atheism and materialism. The first condition of a philosophy capable of leading the world is to establish an absolute and God-centered value system. I see that the philosophy best equipped to lead future world civilization is the Unification Thought of the Reverend Sun Myung Moon. Unification Thought is a new system of philosophy that realizes proper order in the relationship amongst God, human beings, and nature. In this way, we come to discover the true position of God, the meaning of human existence, the meaning of human relationships, and the basis for a harmonious order between humankind and nature.

In his keynote speech at the "International Rally for Victory-Over-Communism and Security" held in Korea on December 16, 1985, the Reverend Dr. Moon described the potential of Unification Thought:

> Unification Thought is a powerful key capable of solving any problem, no matter how difficult it may be. When this thought is applied to society, various social problems can be settled. When this Thought is applied to the world, world problems can be realistically solved. And particularly, when it is applied to Communist ideology and the theory of evolution, all the contradictions of Communism and Darwinism are brought to light, and a counterproposal can be brought forth. This Thought presents a new view of life, a new view of the world, a new view of the universe and a new view of God's work in history. It also offers a principle of integration that can bring different religious doc-

trines and philosophies into unity, while preserving their diverse characteristics.

The Unification Thought Institute (UTI) is dedicated to researching the theoretical foundations needed for a new culture of peace. For he forty years since its founding. since its founding on August 20, 1972 in Seoul, Korea, UTI has sought to address the philosophical and social dilemmas that have led to the chaos and breakdown that undermine modern society. The Institute regularly conducts international seminars and symposia where UT scholars introduce the new thought system and work with other interested parties to unify values across academic disciplines that resonate with values that the world's religions and cultures have long held in high esteem.

Unification Thought was advocated by Reverend Dr. Moon, who completed his earthly life and mission on September 3, 2012. This collection of UT textbooks is meant to provide direction for each of the academic fields introduced in this series. These books are published in commemoration of the Foundation Day of "Cheon-Il-Guk" (the Kingdom of Peace and Unity in Heaven and on Earth) February 22, 2013 (January 13, 2013 by the Heavenly Calendar). On this singular occasion in human history, it is a special privilege for the Unification Thought Institute to dedicate these books to Reverend Sun Myung Moon and Mrs. Hak Ja Han, who, through the Will of God and their unprecedented response to it, now stand as the True Parents of humankind.

Sung-Bae Jin, Ph.D.
President of UTI, Korea

CONTENTS

PREFACE

When I first encountered Reverend Moon's revolutionary insights into the nature of God and His creation while vacationing in California, the idealistic side of my character—which had a brief flirtation with Marxism while studying biochemistry at university—was so ignited and inspired that I was convinced that it would only take a few months, at most, for the world to be transformed.

Unfortunately, this was not the case. It has now been almost forty years and the darkness is still all about us. Reverend Moon has relocated to spirit world and, just yesterday, this misguided idealism of North Korea, his natal country, just successfully launched a ballistic missile capable of carrying a nuclear bomb to the USA.

While the content of the Divine Principle is mainly focused on the religious aspect, the Principle of Creation does have implications for the physical sciences. Rather unexpectedly, at least for the nonbeliever, many of these implications are already entering the mainstream of modern science. I shall just mention a few of these points in this introduction:

1. The Universe is run by the Logos (natural law) that was specifically designed by God with human beings in mind

The concept of natural law was implicit in pre-1900 classical science even if there was no explicit explanation of how an abstract law could influence the behavior of physical matter. This mysterious connection is evoked in Eugene Wigner's legendary lecture on "The Unreasonable Effectiveness of Mathematics in the Natural Sciences."

As science has progressed and the laws expressed to an ever greater precision, it has become increasingly evident that even a minuscule change in the natural laws would not allow for a universe congenial to living organisms. A universe in which the cosmological constant was fractionally different, for instance, would either not expand at all after the Big Bang or expand so rapidly that stars could never have formed. A fractional change in the relative strengths of gravity and electromagnetism would not allow for stars that burn sedately for billions of years. The quadruple-coincidence of specific energy levels in deuterium, helium, carbon and oxygen is essential if the elements of life are to be created by thermonuclear reactions (as an illuminating aside, Agatha Christie, the author of my favorite detective stories, stated that she only allowed one coincidence per story as readers would not stand for more of them.)

As the number of these anthropic coincidences have accumulated they have required an explanation. I favor that of a creative designer. The only other available is the Mutiverse Theory which states that our visible universe is just the one out of a trillion [repeat forty times] trillion trillion (10500) other universes in which the random laws just happen to favor life.

As not a one of this multitude is even theoretically observable, the intellectual choice is between and invisible creator God or an invisible plethora of random universes.

2. *Our Universe had an Origin and Development*

While the Abrahamic theologies always considered the universe to have a creator and thus an origin, much of classical science assumed that the universe was eternal and unchanging. It is only in the last half-century that is become firmly established that the universe had an origin, the Big Bang, and has gone through a series of changes over the last 13 billion years.

It is thought-provoking that while the origin was a complex affair, the Biblical account is quite accurate in that what emerged within the first few minutes of the Big Bang was essentially light, in the form of high-energy gamma rays, with matter as only a one-in-a-hundred-billionth part 'impurity.'

3. *Physical matter has an objective external form—the observable substantial body— and a subjective internal character—the insubstantial mind.*

Classical science, which is still taught in high school and is the science that most people are aware of, was based on the concept that the external, observables of matter are all there is that needs to be taken into consideration.

When science became capable of a more detailed approach, experiment insisted that this simple concept be replaced by the more sophisticated concepts of quantum mechanics. Matter is now understood to have two aspects: an external 'particle' aspect that can be described by conventional mathematics, and an internal 'wave' aspect that can only be described by sophisticated 'complex' mathematics. The internal is subjective and is determined by natural law, the external is objective and faithfully reflects the internal over a adequate time period.

4. The physical universe is just one of the two substantial realms that comprise the cosmos

A science progressed in its attempt to reduce the complexity of the world to the current simplicity of just two types of fundamental entities—the fermion 'bits of matter' and boson 'bits of force' the assumption was that this comprised all that exists in the cosmos, albeit in vast numbers and complicated relationships.

It was therefore a considerable shock just a few decades ago when astronomy discovered that the familiar elements composed of quarks and electrons were only a minor component of the galaxies that stipple the vast emptiness of the vacuum. There was also a major, if invisible, component, named 'dark matter' whose nature is still uncertain but retains the heretofore universal characteristic of mass-energy in that it has a positive gravitational influence (which was how it came to the attention of astronomers).

This blow to scientific complacency was, however, just a quiver compared to the earthquake of recent years when astronomers were forced to conclude that matter itself—both regular and dark—was but a minor component in the cosmos, that 70% of the universe was 'dark energy' that was also invisible but had a negative anti-gravity that was pushing the universe to increase its rate of expansion.

It is now clear that the composition of the cosmos is 70% this not-matter not-energy, 25% invisible dark matter and only 5% familiar 'baryonic' matter.

It is yet to be determined if this major component of the universe—which is not one of the putative multiverses aforementioned—is quantized or if it has a structure to it. But it has imposed on science an unfamiliar humility best expressed in the phrase, "Not only is reality stranger than you think, it is stranger than you can think."

5. Darwinism is incomplete: Evolution by random chance-and-accident is wrong.

The terms 'darwinism' and 'evolution' cover a plethora of concepts, most of which are in accord with Unification Thought. These include:

• All living systems are descended from a common ancestor (prokaryote) which itself emerged from the non-living realm

• The history of living systems stretches over four billion years back to just after the origin of the abiotic earth

• All species, including the human, emerge in history on the foundation of a parental species

• Those entities that are in harmony with the environment as it develops according to natural law will thrive and prosper; those entities that are not in harmony with natural law will eventually wither and decline

There are only two key points where Unificationism differs from classical Darwinism:

1. There is no forethought in evolutionary development

In classical science, there is little concept of natural law beyond the simple rules of physics. Unificationism, to the contrary, has the concept of the Logos which is an extensive construct that has the emergence of human beings as a long-term goal. It is only the Logos that can explain the otherwise unexplainable, but necessary, complementary properties found throughout the living realm. Examples abound from the simple, such as right-handed sugars paired with left-handed amino acids, to the analog manipulations of proteins paired with the digital manipulations of nucleic acids, to the short-term memory of neurons paired with the long-term memory of glial cells in the brain.

2. The changes in a parental species that lead to the emergence of a daughter species are random chance-and-accident

Unlike earlier generations, we are familiar with the conversion of digital information—be it from CDs, DVDs, and hard drives; or as WWW access via cable and WiFi—into the analog form of picture and sound we can easily relate to.

The simple genetics that developed in the second half of the 20th century that was merged in the Modern Synthesis with Darwinism had as a basic axiom, called the Fundamental Dogma, that the flow of information from digital nucleic acid to analog protein was one-way only. From deep-time storage in DNA to short-term storage in RNA to the primary structure of protein to the activity and function of the folded molecule.

This 'read only' situation did not allow for the writing of digital information which, in turn, implied that any change in the digital store could only be by random change, by mutation.

As any content-provider could tell you, this situation is utterly anathema to the creative world of digital movies, music and computer programs. Bill Gates is probably constantly thankful that Windows XP or Microsoft Word is not going to appear given even billions of years of random changes.

Fortunately, modern genetics has just recently become aware that living systems are a fully capable of writing digital information to complement the well-understood capacity to read digital information. This is the emerging science called epigenetics.

In tandem, it is becoming clear that DNA, which has heretofore played the central starring role in biological theory, has only a supporting—if essential—role comparable to that of the somewhat passive role of a hard drive in a PC or Mac computer. All the really important action—including getting life started in the first place—is performed by RNA with a role comparable to the CPU of a computer that reads and writes digital information to the hard drive as well as manipulating it, combing it with input from the environment (the User) before outputting it as analog information to screen or loudspeaker.

For some time, it has been known that RNA information is easily written into DNA information by the activity of Reverse Transcriptase enzyme (RT), but this was only thought to be significant for RNA 'retroviruses' such as HIV. This insignificant, if vexing, role for RT in living systems made it difficult to account for the 500,000 or so different variants of RT found in the human genome. Richard Dawkins discounted this as yet more Junk DNA, but even this concept is dustbin-bound as this DNA that is never translated into protein structure is now known to be transcribed into RNA at low, but significant rates.

Almost every week and new type of RNA is discovered with a new set of properties to add to the dozens of RNA types already uncovered. This is a list of some of the recently established varieties of RNA. Only the first three (unshaded) are included in the classical 'read-only' worldview of random chance-and-accident Darwinism as eloquently propagated by Richard Dawkins et al.

mRNA	Codes for protein	srpRNA	Membrane integration
rRNA	Translates mRNA	snoRNA	Base modification
tRNA	Link to aminoacids	smyRNA	mRNA splicing
miRNA	Gene regulation	teloRNA	Telomeres on DNA
piRNA	Chromosome stability	siRNA	Gene regulation
gRNA	mRNA modification	xistRNA	Chromosome inactivation
rnpRNA	tRNA maturation	aRNA	mRNA translation
yRNA	DNA replication	lncRNA	various

It has been known for some time that RNA is responsible for many central processes in living systems, such as the 'peptidyl transfer' of linking amino acids into proteins. This output activity is complemented by the recently discovered input activity of RNA 'riboswitches' in response to environmental changes:

Riboswitches bind cellular metabolites and control gene expression: Segments of RNA, typically embedded within the 5'-untranslated region of a vast number of mRNA molecules, have a profound effect on gene expression through a previously-undiscovered mechanism that does not involve the participation of proteins. In many cases, riboswitches change their folded structure in response to environmental conditions (e.g. ambient temperature or concentrations of specific metabolites), and the structural change controls the translation or stability of the mRNA in which the riboswitch is embedded. In this way, gene expression can be dramatically regulated at the post-transcriptional level.[1]

In conclusion, while the Divine Principle can lead us to a deep and profound personal experience of how great is the love of God for us, His children; Unification science can lead us to an equally profound appreciation of the genius intellect responsible for such micro-marvels as the ribosome and macro-marvels as the evolutionary history of which we are but a result.

---

1 Roth, A.; Breaker, R. R. (2009). "The Structural and Functional Diversity of Metabolite-Binding Riboswitches". Annual Review of Biochemistry 78: 305–334.

INTRODUCTION

This work is a presentation of the author's concept of a scientific worldview based on the Divine Principle (theological) and Unification Thought (philosophical) worldview as first taught by the Reverend Sun Myung Moon.

This worldview embraces three great realms that, together, make up the Cosmos:

1. The Abstract Realm. This is where God, the Natural Laws and the Mind reside.

2. The Physical Realm. The objective reality in which we have a physical mind and body. We are born, we grow up, and we die.

3. The Spiritual Realm. The objective reality in which we have the spiritual mind and body we developed while in the physical realm. We exist here without end.

Current scientific understanding of the Abstract Realm is restricted to mathematics. The exploration of the Physical Realm is well developed, and will be the main focus of this work. Current scientific understanding of the Spiritual Realm is almost zero although, as we shall see later, a start might just have been made in exploring this realm.

It is not our intention to prove current science 'wrong.' It is, however, our intention to show that current science is incomplete; that it is a subset of a larger picture. The Unification view as presented here has four important implications for current science:

That spacetime is a two-sided entity, and that the physical realm only inhabits only one side of it.

That natural law is a sophisticated, not simple, abstract construct that is mathematically-similar to spacetime.

That it is incorrect to ignore the quantum wavefunction aspect of physical systems in all the sciences except physics and chemistry.

That living systems involve the RNA-controlled interplay of analog form and digital memory. Their evolution history is driven by accumulated wisdom, not by chance-and accident.

Each of these four points will be explored in detail as the discussion progresses.

Science and religion take the opposite tack in their attempts to explain the world we live in. Religion is a 'top-down, big-picture' discipline; it starts with God and works downwards to dealing with simpler matters.

Science is a bottom-up discipline; it starts at the bottom with the simplest things, and works upward towards comprehending the big-picture.

This work takes the scientific approach, so God will only appear towards the end of the discussion, not at the start. We will find it necessary, however, to replace the vague and simple concept of Natural Law with a much more sophisticated and mathematically well-defined entity we will call the Logos. Only towards the end of this work will we need to discuss how and why such a sophisticated abstract entity as the Logos came into existence. Only the natural law Logos which is immutable and unchanging will be necessary to understand the structure and history of the physical realm up to the origin of Man. We will not have to invoke any direct action of God, a view that is compatible with religious history and the obvious immense difficulty that God has in directly influencing events in the physical realm. In the final section we will briefly discuss the reasons for such a 'hands-off' approach to the creation and running of the physical realm.

We start with the very basic question of all: Why is there 'something' rather than 'nothing' and why we have to include an Abstract Realm if we want to aspire to a complete picture of the 'something' we find ourselves in.

1.
THE ABSTRACT REALM

While there is probably a technical term in philosophy for the simplest of entities that inhabit the Abstract Realm, here I shall just use the term 'concept.'

We start with two concepts: that of 'absolutely nothing' and 'a set.'

The concept of a set and what it contains is a fundamental concept. It is commonplace in language where the word 'humans' is the set of all humans, 'cows' the set of all cows, 'integers' is the set of all counting numbers, etc. It could also be a set of simpler sets.

The contents of a set are usually placed inside two curly brackets and separated by commas; so the word 'cows' is the set {cow1, cow2... cowN} while the integers are the set {1, 2, 3, ...}—the first being a finite set while the second is an unbounded set as there is always a larger integer.

LINEAR EXTENSION

'Absolutely nothing' is the concept we call the empty set, the null set and its symbol is {} and it is said to have zero, 0, members. Any set whose members can be put in a one-to-one correspondence (1to1) with this set is said to be of count, or size, zero, the integer (the counting numbers) symbolized by 0.

This null set is now made the member of a second set, a set with a single member, the null set. Any set that can be put into 1to1 with this is said to have one member, the integer 1.

The concept of absolutely nothing automatically leads to the concept of one, so we no longer have 'absolutely nothing', we automatically generate the concept 'one.'

$$\{\ \} \equiv 0$$
$$\{\{\ \}\} \equiv 1$$
$$\{\{\{\ \}\}\{\{\ \}\}\} \equiv 2$$
$$\{\{\{\ \}\}\{\{\ \}\}\{\ \}\} \equiv 3$$

As there is an unending supply of nothing we can make endless number of these sets of 1 and by combining them generate all the integers, one after the other in linear order, starting with 0 and going on without bound—which is often simply illustrated on the integer line, starting off at zero and stretching off to an infinity of integers.

The integers

This is the additive concept that generates the integers. Now while the structure of the abstract realm is vertically hierarchical, there are also many horizontal connections. The 'power set' is another approach to getting something, albeit an abstract concept, out of the concept of Absolutely Nothing.

Given a set of members, how many unique unordered subsets, called the power set, can be constructed. As each member of the set can either be in a subset or not in a subset, there are two possibilities for each member. The number of subsets is thus easily calculated by multiplying by 2 for each member. The power set of a three-membered set is going to be 2 x 2 x 2, symbolized by 23, which is eight.

You can always make two 'improper' subsets by either taking none of the members, the null set, or taking all of the members, the entire set. All the 'proper' subsets have at least one member in them. A simple set of three members: A,

$$\{A,B,C\} \xrightarrow{\text{power set}} \left. \begin{cases} \{\ \},\{A,B,C\} & \text{\textit{improper subsets}} \\ \{A\},\{B\},\{C\} & \text{\textit{proper subsets}} \\ \{AB\},\{AC\},\{BC\} \end{cases} \right.$$

B and C will serve to illustrate. Note that only different sets, ignoring ordering are considered, i.e. {B,C} = {C,B} so only counts once.

We start with the power set of the set with just one member, the null set, the first to appear out of nothing. We can either take this set or we cannot take this set. In either case we end up with the empty set. As we are only interested in different subsets when calculating the power set this is just one member of this power set. The power set of a zero-member set is measured by the integer 1, a single set. This gives us 20=1, an 'exponential' expression that has caused much confusion.

Now the 'set whose one member is the empty set' is not the same as the empty set, so the power set of a one-member is measured by the integer 2, and in exponential notation, 21=2.

Next we take the power set of 2, and this has four members, 22=4. We can iterate this process and create a hierarchy of power sets. These rapidly get very large: the power set of a 65,000-member set is an integer that is 19,000 digits long—just printing it here would take up all the room in this book.

A more gradual hierarchy of exponentials is given by the simple rule of adding exponents while multiplying. Just by iterating 2N+1=2×2N we create each level in turn.

$$2^2 = 2 \times 2$$
$$2^3 = 2 \times 2 \times 2$$
$$2^2 \times 2^3 = (2 \times 2) \times (2 \times 2 \times 2)$$
$$= 2^{2+3} = 2^5$$

Using this hierarchy we have the basis for the binary expression of the integers rather than our familiar decimal system involving adding powers of ten together, where the integer 132=1×102 + 3×101 + 2×100 in decimal and 10000100 in binary.

Even in the temperate binary sequence the size of the power set grows very quickly: for a ten-member set, the power set already has over 1,000 members. Gregor Cantor was the first to prove that for any set, be it finite or infinite, the power set is always has more members, it cannot be 1to1 with the it. This is a trivial observation for finite sets, but not so obvious when dealing with infinite sets, and this leads to some new insights into infinity.

We have already encountered infinity in the unbounded integers. This is called the countable infinity, and as it is the first of many, and this countable 'denumerable' infinity is called 'aleph-zero.' A simple but unintuitive fact is that the infinite set of even numbers can be put into a 1to1 with the set of integers (which intuition suggests that as this has all the odd as well all the even numbers, should be twice as big.

$$\text{countable infinity} = \aleph_0$$

1	2	3	4	...	N	...	$\aleph_0$
↓	↓	↓	↓	↓	↓	↓	↓
2	4	6	8		$2N$	...	$\aleph_0$

This simple example illustrates how different infinity is to an integer in which multiplying by two always makes a greater integer while $2 \times \aleph_0 = \aleph_0$, and that a proper infinite subset can be the same size as an infinite set.

Passive and active

Starting with the concept of absolutely nothing, we inevitably ended up with the infinite integers. Next we have the concept of integers interacting with each other. The operation of addition is akin to placing two integers side by side and seeing what new integer they are equal to. In this 'placing them together' operation the integers play a passive role. Subtraction, the reverse of addition, leads to the negative integers which zoom off from zero in the opposite direction to the counting number line to negative infinity, and we will discuss them in more detail in a moment.

In multiplication, one of the integers plays an active role while the other plays a passive role. This difference can be seen in the sensible "five times four oranges" versus the nonsense of "five oranges times four oranges." The active role is "five" and the passive role is "four oranges." The active integer 'five' is stretching the passive integer 'four' and transforming it into the integer 'twenty.'

These two roles are not so apparent in the integers where multiplication is 'commutative' and the order is irrelevant. There are, however, important areas of mathematics which are distinctly non-commutative and the order of multiplication is crucial.

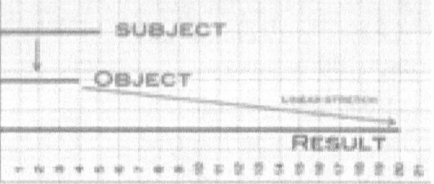

$$x \times y = y \times x \quad \text{commutative}$$

$$x \times y \neq y \times x \quad \text{noncommutative}$$

The squaring function is when an integer acts on itself; it takes both the active and the passive role.

The inverse of multiplication is an integer divided by another integer (except zero) which leads to the fractions, the 'rational numbers.' There are a countable infinity of these as can be seen in an array

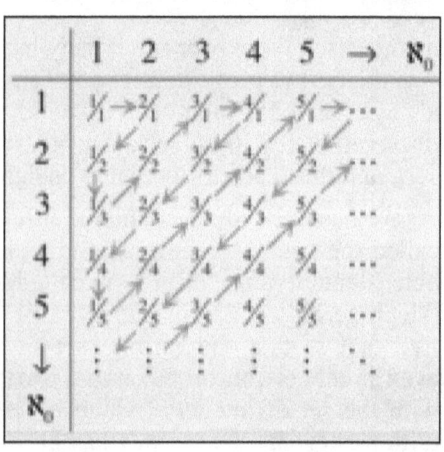

where the ratios of all the integers is to be found. Starting at the corner with 1/1, we can now count all of them in order along the diagonals (that some ratios, such as 1/2 and 2/4, are identical is irrelevant) with a complete 1to1 correspondence of the integers to the array of rational numbers. So we have

$$\aleph_0 = \aleph_0 + \aleph_0$$

$$= \aleph_0 \times \aleph_0$$

Many infinite sets are countable and can be put into a 1to1 correspondence with the integers. It was a great surprise to the early Greeks to discover that there were numbers along the number line that could not be expressed as a ratio of two integers. A particularly simple example is the length of the diagonal of a square of one unit. Now Pythagoras had proved that in a flat geometry the square of the hypotenuse is equal to the sum of the squares of the other two sides, so the length of the diagonal, d, squared was equal to the sum of one squared plus one squared.

The number, d, when squared resulted in the number 2, so d is the square root of 2.

It did not take long before it was realized that there is no ratio of two integers that when squared, results in the integer two. The diagonal was not measured by a rational number, it was a number that was 'irrational.' A modern approach to proving this irrationality is used in the final expression: that there is no integer that when squared is equal to twice another integer squared. This approach involves another concept that emerges naturally when integers are multiplied and divided; the concept of the prime numbers.

$$d^2 = 1^2 + 1^2$$
$$= 2$$
$$d = \sqrt{2}$$
$$\neq \frac{n}{m}$$
$$2 \neq \left(\frac{n}{m}\right)^2$$
$$2m^2 \neq n^2$$

Primes

Integers fall into three classes in terms of their behavior in multiplication and division.

The Defining integers. There are just two of these, 0 and 1. Multiplication by 1 does nothing at all, nothing changes; while division by 1 also has no effect. Multiplication of any integer, no matter how huge, by zero results in zero; while division of any integer by zero does not result in an integer (NAN=Not A Number) but an unbounded, endless process heading off to infinity. This is why division by zero is not allowed.

$$1 \times n = n$$
$$\frac{n}{1} = n$$
$$0 \times n = 0$$
$$\frac{n}{0} = \text{NAN}$$

The Prime integers. A prime number, p, can only be created by multiplying the prime by the unit, and it cannot be divided by another integer to give another integer.

$$1 \times p = p$$
$$\frac{p}{m} \neq n$$

The Composite integers. A composite number, c, can be created by multiplying a set of primes together, its 'prime factors,' and it can be evenly divided by these primes.

$$c = p \times q$$
$$\frac{c}{q} = p$$

The 'fundamental theorem of arithmetic' says that every positive integer has a unique prime factorization. Although the proportion of primes falls off as the integers get larger, it was proved by the early Greeks that they are infinite in number; there is no largest prime.

The first few prime numbers are 2, 3, 5, 7, 11, 13, 17, ... All integers, except the defining integers 0 and 1, have a unique set of prime factors, and this is a listing of 2-16 and 102-116 (with the primes indicated in red.

$2 = \{2\}$	$102 = \{2,3,17\}$
$3 = \{3\}$	$103 = \{103\}$
$4 = \{2,2\}$	$104 = \{2,2,2,13\}$
$5 = \{5\}$	$105 = \{3,5,7\}$
$6 = \{2,3\}$	$106 = \{2,53\}$
$7 = \{7\}$	$107 = \{107\}$
$8 = \{2,2,2\}$	$108 = \{2,2,3,3,3\}$
$9 = \{3,3\}$	$109 = \{109\}$
$10 = \{2,5\}$	$110 = \{2,5,11\}$
$11 = \{11\}$	$111 = \{3,37\}$
$12 = \{2,2,3\}$	$112 = \{2,2,2,2,7\}$
$13 = \{13\}$	$113 = \{13\}$
$14 = \{2,7\}$	$114 = \{2,3,19\}$
$15 = \{3,5\}$	$115 = \{5,23\}$
$16 = \{2,2,2,2\}$	$116 = \{2,2,29\}$

$$n = \{p^a q^b\}$$
$$n^2 = \{p^a q^b p^a q^b\}$$
$$= \{p^{2a} q^{2b}\}$$
$$15 = \{2^0 3^1 5^1\}$$
$$15^2 = \{2^{2\times 0} 3^{2\times 1} 5^{2\times 1}\}$$
$$225 = 1 \times 3 \times 3 \times 5 \times 5$$
$$2 \times 15 = \{2^0 3^1 5^1\}$$
$$2 \times 15^2 = \{2^1 3^2 5^2\}$$
$$450 = 2 \times 3 \times 3 \times 5 \times 5$$

$$d^2 = 1^2 + 1^2$$
$$d = \sqrt{2}$$
$$\sqrt{2} = {n}/{m}$$
$$2 = {n^2}/{m^2}$$
$$2m^2 = n^2$$

odd power of 2 = even pow

When an integer is squared, the number of primes in the result is just double that of the integer. It is simple to see, in exponential notation, that the prime factors of an integer-squared must come in even numbers. As zero is an even integer—it has a zero remainder when divided by 2—this also holds for primes that are not in the factorization, such as 2 in the integer 15.

In particular regard to the existence of irrational numbers, this implies that the result of squaring any integer, n, will be an integer with an even power of 2s in its prime factorization. Multiplying the resultant integer by 2 will result in an integer with an odd number 2s in its prime factorization, as in the example.

Returning to the length of the diagonal of a unit square, assuming that it is the ratio of two integers leads to equating two integers, one with an even, and the other with an odd, power of 2s in its prime factorization. As this is not conceivable, the diagonal cannot be the ratio of two integers and it must have an irrational length. (The name given them only suggests the extreme discomfort the Greeks experienced by this discovery.)

Using this method, all the square roots of the integers that have an odd-power prime among their prime factors (which is trivially the case of the primes) can be shown to be irrational numbers.

Infinity of infinities

We have already noted that there are a countable infinity of fractions. We can, however, generate a countable infinity of the 'proper' fractions that lie between zero and unity by the reciprocal function, one divided by each of the integers.

The reciprocal function can pack all the integers from 1 out to the unbounded infinity of integers in order into the space between 0 and 1 without missing a single one. Just as infinity is not a number but a limit, so is zero in division (although it is treated as an honorary integer, it's not really, as hinted at by the edict against 'division by zero'). This is an emergent property of integers that are 'interacting' by division, the inverse of the multiply operation. And there is plenty of room for the fractions such as ⅔ that are not on this list. This is an illustration with a vengeance of the fact that a countably infinite set (all the ratios of integers) can have a countably infinite number of infinite proper subsets (the fractions between 0 &1, the fractions between 1& 2, the fractions between 2 & 3, etc.)

The rational numbers are everywhere dense, but somehow there are gaps between them into which the irrationals fit along the number line.

All rational numbers, both integers and fractions, can be expressed as decimals that have a repeating pattern (which can be all zeros) off to infinity (which is usually indicated by a bar over the repeat). Irrational numbers do not have any repeating pattern to them.

$\frac{1}{1}$	$1.00000000000\overline{0}$	rational
$\frac{1}{2}$	$0.50000000000\overline{0}$	
$\frac{1}{3}$	$0.33333333333\overline{3}$	
$\frac{1}{7}$	$0.1428571\overline{42857}$	
$\frac{2}{11}$	$0.1818181818\overline{18}$	
$\sqrt{2}$	$1.4142135623730...$	*irrational*
$\sqrt{3}$	$1.7320508075688...$	

It took almost 2,000 years before Georg Cantor proved that the infinity of irrationals was greater than the countable infinity of the rationals. The infinity of irrational decimals was not countable, it was the first example of an 'uncountable' infinity. He showed, by an ingenious diagonal argument involving choosing its nth digit to be different from the nth digit of the nth decimal, that it was always possible to construct a decimal that was not on any list. The irrationals could not be put in a 1to1 matching with the integers, there were always irrationals left over.

The counterintuitive conclusion is that, even though the rational numbers are every-where dense along the number line, they leave space for a much greater infinity of irrational numbers along the continu-ous number line.

Cantor also showed that the power set of any set, be it finite or infinite, is always greater than the set itself. (A trivial concept for finite sets, but of great subtlety when dealing with infinite sets.) He thus established a hier-archy of infinities founded on the countable infinity of the integers, which he called 'aleph zero.' The power set of aleph-zero and the uncountable infinity of the irrationals is this 'infinity of the continuum' called aleph-one. The power set of aleph-one is aleph-two, an even greater infinity. The taking of power sets can be continued without bound, resulting in a countably-infinite hierarchy of uncountable (except for the first) infinities each greater than the previous.

$$1,2,3,4\ldots = \aleph_0 \text{ rationals}$$
$$2^{\aleph_0} = \aleph_1 \text{ irrationals}$$
$$2^{\aleph_1} = \aleph_2 \text{ shapes}$$
$$2^{\aleph_n} = \aleph_{n+1}$$
$$\ldots = 2^{\aleph_{\aleph_0}}$$

This is quite something, starting with nothing we inevitably end up with the concept of an infinity of infinities. This is the nature of the Abstract Realm as higher levels with more and more sophisticated emergent properties emerge from lower levels without those properties.

Transcendental numbers

To complete our list of numbers along the linear number line we have to accept that some important irrational numbers are more irrational than others; they are transcendentally irrational compared to the algebraically irrational numbers just discussed. There are actually two classes of irrational numbers, numbers that cannot be expressed as the ratio of two integers or as a repeating decimal.

Algebraic irrationals. These involve a finite set of operations on the integers—a finite number of additions, multiplications, taking roots, etc. More precisely, mathematicians say "are roots of a finite polynomial with integer coefficients." So $\sqrt{2}$ is an algebraic irrational; a single operation is involved, and it is a one root, or solution ($-\sqrt{2}$ is the other) to the quadratic polynomial, $x^2 - 2 = 0$

Transcendental irrationals. These involve an infinite number of operations on the integers. They cannot be expressed in a finite form.

We will mention just two of the important transcendental numbers that play important roles in the structure of the Abstract Realm, one is called 'pi' and the other just 'e'.

The number π (pi) can only be equated with an infinite number of operations. One such awkward looking equation involves adding up six times the reciprocals of all the integers-squared and taking the square root of the result.

$$\pi = \sqrt{\lim_{n \to +\infty} \frac{6}{1^2} + \frac{6}{2^2} + \frac{6}{3^2} \cdots \frac{6}{n^2}}$$
$$= 3.14159265358979323846264\ldots$$

While π appears in many unexpected places in the mathematical structure, its history is not so much connected with linear extension as with circular rotation (a topic we will soon deal with in detail).

The diameter of a unit circle (one with a radius of 1) cuts the circle into two halves that each have a length of π (while the straight line distance is $2\sqrt{2}$), while the area of the unit circle is also π.

A spot in circular motion on a circle of linear radius 1 making one 'period' travels a distance of 2π going once around the circle. This measure of rotation is called radians, so there are π radians in ½ a period, and ½π radians in ¼ of a period.

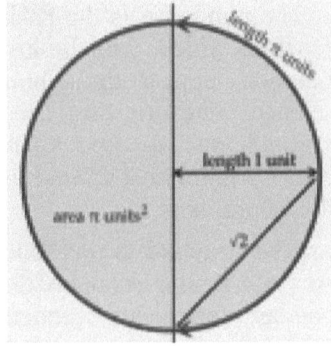

While π gives a relation between linear and angular motion, the number 'e'—the 'base of natural logarithms,'—deals with the balance between the unit and the infinite.

If you multiply the unit 1 an infinite number of times it remains 1. If you multiply 2 in this way, the result rapidly zooms off towards infinity.

For numbers only slightly larger than 1, the result depends on how many times you repeat the multiplication. Enough times, and any number even infinitesimally greater than 1 will eventually zoom off to infinity. If the number of repetitions is small, however, the result stays close to 1. As an example, we will use a number, n, that is just one millionth greater than 1. When multiplied just a hundred times, the result is also just barely greater than 1. Multiplied a hundred million times and the huge number that results has 43 digits before the decimal point.

$$e = \xrightarrow[n\to\infty]{\lim}\left(1+\frac{1}{n}\right)^n$$

$$n = \left(1+\frac{1}{1,000,000}\right)$$

$$n^{100} = 1.000100005$$

$$n^{100,000,000} = 2.6\times10^{43}$$

$$n^{1,000,000} = 2.718280469$$

$$e = 2.718281828...$$

There is a balance point between these two extremes, a Goldilocks point of being 'just right' we might say. If the number is raised to the power, n, that is the reciprocal of just how much the number is greater than $1+1/n$, the result in the infinite limit ends up as the number 'e', which is between 2.5 and 3.0.

The number e is also the limit of adding an infinite number of fractions. The partial series with just 7 fractions to add yields e correct to three decimals. Both of these formulas are a lot more elegant than the one for pi.

$$e = \sum_{n=0}^{\infty}\frac{1}{n!}$$

$$= \frac{1}{0!}+\frac{1}{1!}+\frac{1}{2!}+\frac{1}{3!}+\frac{1}{4!}+\frac{1}{5!}+\frac{1}{6!}\cdots$$

$$= \frac{1}{1}+\frac{1}{1}+\frac{1}{2}+\frac{1}{6}+\frac{1}{24}+\frac{1}{120}+\frac{1}{720}\cdots$$

$$= 2.71805555$$

$$e^x = \sum_{n=0}^{\infty}\frac{x^n}{n!}$$

Many operations in mathematics involve exponential behavior (ever faster) or logarithmic behavior (ever slower) which are both connected to

e. The number e is the basis for the exponential function which, with its inverse, the natural log function, appear throughout mathematics and science. (The term 'log' used to refer to a base of 10, and 'ln' was used for the natural log to the base e, but in most science and math 'log' implies the natural base.

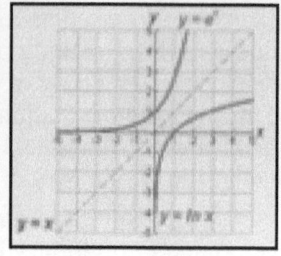

The structure to this foundational level of the Abstract Realm has many unexpected connections, such as the one between the transcendental number 'e' and the prime integers.

$$x = e^{y}$$

$$y = \log x$$

$$x = e^{\log x}$$

Distribution of the primes

Ever since they were discovered, people have been interested in just how the primes are distributed among the integers. The table of a few primes that we presented above illustrates some facts about the primes:

All the primes but the first are odd numbers. As 50% of the integers are even numbers, this eliminates 50% of the integers. Similarly, ⅓ of the integers are divisible by 3, ⅕ of them by 5, 1/7 of them by 7 and so on.

As the integers get larger, the primes get fewer in number.

The spacing between consecutive primes can be as small as 2 and varies in a seemingly random fashion.

In such studies the number of primes equal to, or less than a given number, n, is called the prime counting function and symbolized by π(n), an example of overloading a symbol as the π function is not the same as the number π.

The table lists π(n) up to 11 (note the plateau between 7 and 11) and the graph gives its value for the first 140 integers.

$$\pi(2) = 1$$
$$\pi(3) = 2$$
$$\pi(4) = 2$$
$$\pi(5) = 3$$
$$\pi(6) = 3$$
$$\pi(7) = 4$$
$$\pi(8) = 4$$
$$\pi(9) = 4$$
$$\pi(10) = 4$$
$$\pi(11) = 5$$

Just a few centuries ago, the proof of the Prime Number Theorem was established. It is this PNT theorem that connects the transcendental 'e' with the integer primes. The PNT states that the number of primes less than a number, n, is approximately that number divided by its natural logarithm, and

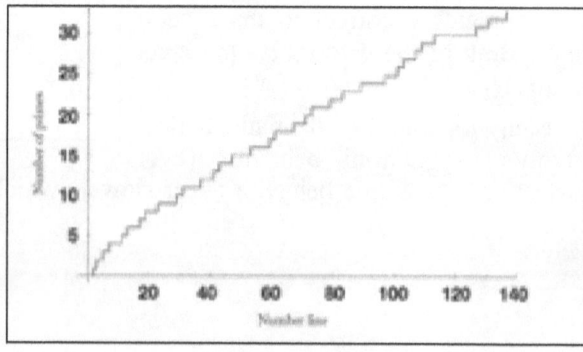

that the relative 'error' in the count falls to zero as the number gets unboundedly large.

$$\pi(n) \approx \frac{n}{\log(n)}$$

It will be seen from the graph of $\pi(n)$ up to 100,000 that while the absolute error is increasing (it is about 1,000 at 100,000) the relative error is decreasing (it is 1,000/100,000 or 0.01).

$$\frac{\pi(n)}{\frac{n}{\log(n)}} = \xrightarrow[n \to \infty]{} 1$$

So there is some order to the appearance of primes amongst the integers. The focus then moved to the error in this estimate. A great improvement involved integrals of the logarithmic function, and currently a complete understanding of the distribution of the primes rests on a proof of the Riemann Hy-

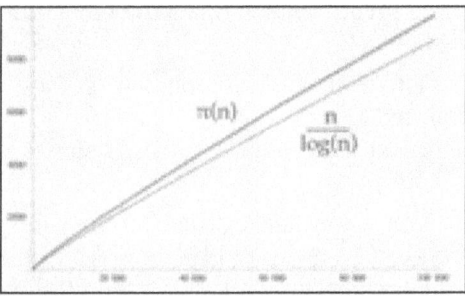

pothesis about the zeta function. We will return to this topic later, but it should be noted that we have already touched on yet another Abstract Realm connection in our formula for the transcendental number π. This formula was just a restatement of a formula for calculating the value of $\zeta(2)$, the zeta function of the integer 2.[2]

$$\zeta(2) = \frac{1}{1^2} + \frac{1}{2^2} + \frac{1}{3^2} + \frac{1}{4^2} + \frac{1}{5^2} \cdots$$
$$= \frac{\pi^2}{6}$$

****

So far, our discussion has been based on the concept of linear extension, or size. The image is that of the integers emerging from zero and extending ever outwards in a line. All the numbers we have discussed so far—integers, fractions, irrationals and transcendentals—are on this linear extension from zero.

Along with this fundamental concept of linear extension, there is another concept that is almost as fundamental, that of angular rotation. While this is also an unbounded, you never get anywhere but just go around and around in circular motion, which is the topic of the next section.

---

2 There are (at least) two excellent introductions to this subject:

Derbyshire, John (2003-04-15). Prime Obsession. Joseph Henry Press.

Rockmore, Dan (2007-12-18). Stalking the Riemann Hypothesis: The Quest to Find the Hidden Law of Prime Numbers. Vintage.

CIRCULAR ROTATION

We have already encountered the unit circle in our discussion of the number π.

We start with a unit extension from zero (linear extension comes first). This establishes the 'real' axis with a rotation of zero. Using the origin as a fixed pivot, we then rotate the line a full circle until it comes back to where it started.

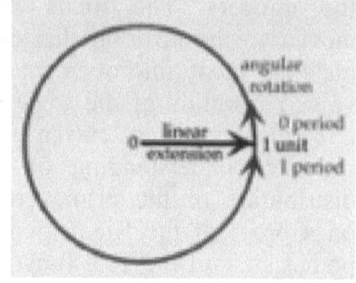

This full circle of rotation is variously referred to as:
1 period = 2π radians = 4 right angles = i4 = 360°.

For a constant circular motion, the period can be measured by time. An example is a circular motion that repeats 60 times a second, it has a frequency of 60/sec and the period is 1/60th of a second.

In all our diagrams you might have noticed that all the circular motion is in an anticlockwise direction. This is the convention for the plus direction that is invariably used in discussing angles. Clockwise is just as possible, and is designated the minus direction.

Circular motion is found throughout the mathematical foundations of the Abstract Realm where it often appears as its linear components called sine and cosine.

Sine and cosine functions

While sines and other such 'trigonometric functions' are usually introduced via static right angle triangles, they are more intuitively grasped when thought of as the connection between circular and linear motion.

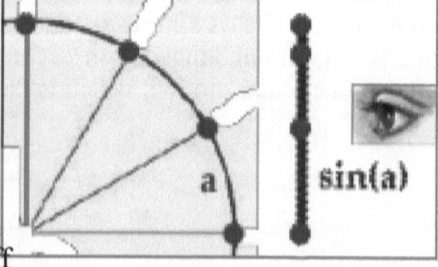

If we look directly at a constant circular motion, the end of the unit extension moves in a full circle, measuring off equal lengths per tick of time. If we look at the circular motion from the side, however, that end appears to just go up and down along a linear line in a movement that does not measure off equal lengths per tick of time. The illustration is for the ¼ of a period, the sine of the angle, a, sin(a). At 0 the sine is zero, it then increases rapidly at a decreasing rate until it slowly arrives at 1 at ¼ period. The reverse happens in the second quadrant as the sine slowly decreases at an increasing rate until it arrives back at 0. In the third quadrant the opposite happens, the sine decreases to −1 at ¾ before returning back to 0 at a full period. This up and down movement repeats for each cycle of circular motion.

This 1-D view of the 2-D circular motion is called the sine motion. It is an example of 'down-sampling'. There are two things to note about the relationship between circular angular motion and sine linear motion:

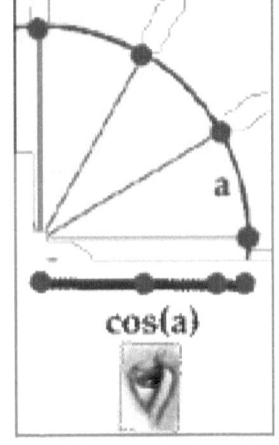

The circular motion is in two dimensions and involves a constant change in position

The sine motion is in one dimension and involves a constantly changing change in position.

While the angle change is in regular steps and increases without limit, the sine change is in irregular steps and oscillates symmetrically between +1 and −1. The sine function compresses the two dimensional motion into a simpler one dimensional component of linear motion.

If we look sideways from the bottom a similar motion occurs that starts at +1 goes to zero at ¼ period, to −1 at ½ period, 0 at ¾ and back to 1 after 1 period. This is the cosine component of circular motion.

The sine and cosine movements are equal and opposite complements. When one is unity, the other is zero, for example, and the two functions are said to be out of phase by 90°. From the Pythagorean relation, the sum of the squares of the two components is unity (as in the first equality, which is invariably symbolized by the second, somewhat misleading, equality).

period a	0	$\frac{1}{4}$	$\frac{1}{2}$	$\frac{3}{4}$	1
sin(a)	0	+1	0	−1	0
cos(a)	+1	0	−1	0	+1

$$[\sin(a)]^2 + [\cos(a)]^2 = 1$$
$$\sin^2(a) + \cos^2(a) = 1$$

If the circular motion that is generating them flips from counterclockwise (plus) to clockwise (minus) it has no effect on the 'symmetric' cosine function but reverses the 'asymmetric' sine function.

$$\cos(-a) = \cos(+a) \quad while \quad \sin(-a) = -\sin(+a)$$

Symmetry

Symmetry plays an important role in the Abstract Realm. A simple example is reflection about a central axis. The letter A, for example, is symmetrical about the vertical center while the letter B is not. The 'B' turns into a 'B' as seen in a mirror. The letter 'A' is said to be symmetric while the 'B' is asymmetric.

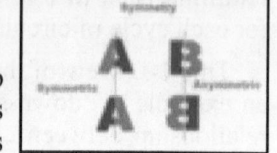

So a single flip does not change a symmetric object. Flipping the letter A a second time also makes no change. This is similar to multiplying plus one by itself repeatedly; at each step you always have plus one.

If you flip the letter B a second time, you get back the original letter B. Flip it again, you have the inverted letter. This is similar to multiplying minus one by itself repeatedly; at each step you alternate between plus one and minus one; it has a period of 2.

Symmetrical, period 1, $(+1)n$ $\quad=\quad +1 \quad +1 \quad +1 \quad +1 \quad +1 \quad +1$

Asymmetrical, period 2, $(-1)n = -1 \quad +1 \quad -1 \quad +1 \quad -1 \quad +1$

This is why a symmetrical object is said to have a positive (or even) symmetry, while an asymmetrical object has a negative (or odd) symmetry.

Mathematical functions also have similar symmetry properties.

We have seen that the sine and cosine functions are identical except for a matter of phase; the cosine is a maximum at 0 or a ½ rotation, while the sine is a maximum at ¼ and ¾ of a turn about the circle.

It is this difference that makes the sine movement an asymmetrical function while the cosine function is a symmetrical function. This can be easily seen if the two curves are drawn on a graph where the zero axis of rotation is at the center and +½ and –½ of a period are placed at either end.

When the cosine (seen in red) flips about the vertical axis it stays exactly the same shape, just like the letter 'A'. The cosine view of circular motion is a symmetrical function.

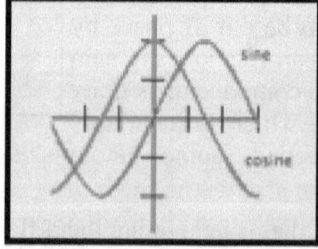 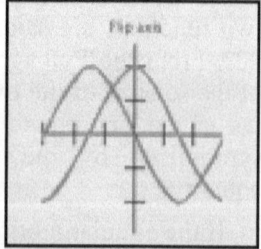

Flipping the sine function turns it into a different shape, the mirror image, just like the letter 'B' — the sine function is an asymmetric function. There is one final difference between the sine and cosine function.

The sine wave has 3 nodes; one at either end and one at the very center, and 2 disconnected crest areas (or antinodes) in lobes on either side of the zero where all the amplitude of the wave is concentrated.

The cosine function is the opposite. Both the cosine and the flip-cosine wave have just 2 nodes where the wave is at 0 (at ±¼ rotations), and all the waviness is concentrated in just 1 antinode.

WAVES

The circular motion is taking place on two of the three dimensions. If the circle now moves along the third dimension, this linear motion combines with the sine motion to generate a sine wave. The distance the wave travels in one period is called the 'wavelength.

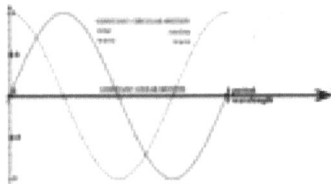

In a similar way, the cosine movement combines with the linear motion to create a cosine waves. The sine wave and the cosine wave are identical except for being 90° out of phase. This phase difference makes the sine wave a 'closed' wave that is zero at either end of the cycle, and the cosine wave an 'open' wave that is at a maximum at either end of the cycle.

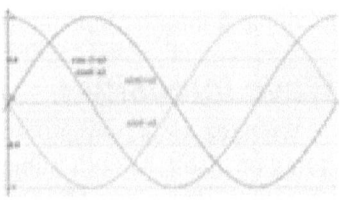

A flip of the circular motion generating the components from anticlockwise to clockwise has no effect on the cosine wave, while the sine wave flips and 'waves' in the complementary direction.

Waves appear throughout the Abstract Realm and the difference between open symmetrical waves and closed asymmetrical waves is often of significance.

The sine and cosine waves we have just discussed are components of circular motion on the unit disk. This creates a sine/cosine wave with an 'amplitude,' or size, that has a maximum of ± 1.

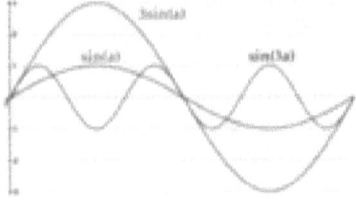

Larger and smaller amplitude waves can be created by just multiplying these 'unit' functions by a scaling factor. For example, the wave $3\sin(a)$ has an amplitude 3 times greater than a $\sin(a)$ wave, while the wave, $\sin(3a)$ has a frequency that is 3 times as great and a period that is ⅓.

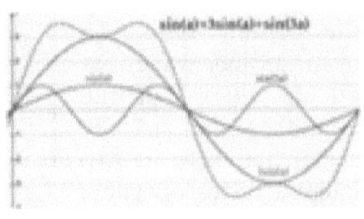

One important property of waves is that two or more of them can seamlessly blend into a single wave. The three waves just mentioned combine into a wave with the unit period with four 'crests' in the amplitude per period. In the 1800s, Joseph Fourier showed how any wave could be broken down into a sum of simple waves, and that any wave, no matter how complex, could be created out of a sum of simpler waves. Some waves, such as the jagged sawtooth wave, are the limit of an infinite sum of waves.

The illustration shows how the abrupt sawtooth wave is the limit of a simple summation of a combination of sine waves of the form 1/n(sin(na)).

The sum when n=1to 3 and n=1 to 10 is illustrated in green, while n=1 to 100 is in red. Even this number of smooth waves is already getting close to the sawtooth wave. In the limit, as n goes to infinity, the sharp sawtooth wave (in blue) emerges. We will now look at a few of the properties of

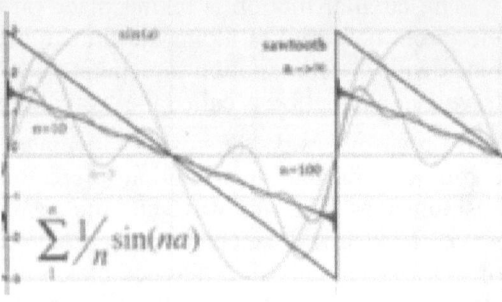

waves that will be important when the discussion moves out of the Abstract Realm and into the Substantial Realms.

Wave intensity

We have already mentioned the squaring function, where a number operates on itself in both subject and object roles. This 'self' action is also important for waves. The square of a wave is called its 'intensity,' and the intensity of the minus half of the wave is equal to that of the positive half, by the rule of signs (which we will deal with shortly) where $-1^2 = +1^2 = +1$.

A flip from anticlockwise to clockwise motion has no effect on the intensity of a sine wave.

The wave intensity of a (co)sine wave does not have the same shape as a sine wave, it is somewhat 'sharper' in shape and varies from 0 to +1. The intensity is a measure of the energy, or power, in a wave which is all in the ±crests and zero at the nodes.

A closed sine wave has zero energy at the boundary and at the center of

the wave. All its energy is in two 'lobes' about the center node. An open cosine wave has all its energy at the boundary and at the center, and there are two internal nodes on either side of the center.

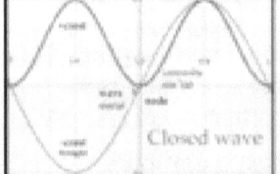
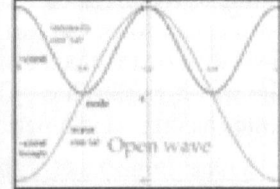

Closed wave   Open wave

Standing waves

We have dealt so far with just single waves. A more sophisticated wave is one where the circular motion continues and the sine and cosine movements cycle endlessly as they pass a reference point.

If such a "traveling wave" has no discernible beginning or end, it is impossible to tell whether it is a sine or a cosine wave. It is also impossible to tell if the related circular motion is clockwise or anticlockwise. The wave can be represented as a simple wave that repeats endlessly.

Unlike the traveling wave that is always changing its location and has global properties, a confined 'standing wave' stays in the same location. It is still an endless wave, but it is reflected endlessly between its boundaries. Standing waves are localized, confined waves.

When dealing with waves and boundaries, an important measure is the wavenumber, the number of waves that fit into the space allotted.

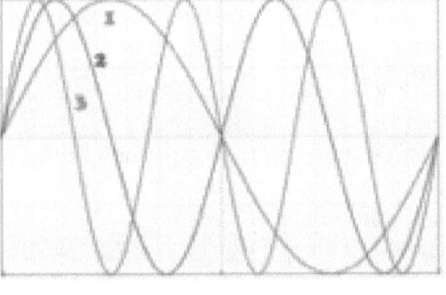

If a sine wave fits by waving once, another will fit by waving twice as will one waving three times. These waves are simply sin1x, sin2x and sin3x, as illustrated, or in general, "sin nx," where n is the integer number of waves that fit. We are now going to change the graph. Instead of going off the right side and appearing on the left, we now imagine that the wave is reflected by either end. Now ½ the period fits into the space.

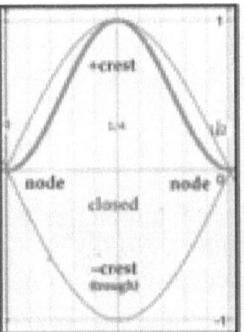
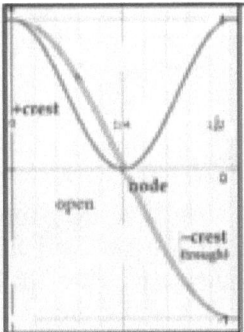

The simplest example is one-half of a sine wave. When it is reflected at the boundary it turns into the other half of the sine wave. All the energy of the bounded wave is in the center of the wave. Examples of closed waves in the Physical Realm are 'closed' organ pipes, the photon of light, and the electron orbitals of atoms.

For an open wave, all the energy is in the boundary of the wave, and there is no energy at the center. Examples are the 'open' organ pipes and the gluons of the strong nuclear force.

For waves that are zero at the boundaries, we have a simple series of waves that will fit—rather like a string vibrating with fixed, zero-movement ends. Their periods are ½, 1, 3/2, 4/2... of the fundamental period, and the numbers in red are the wavenumbers, the counting integers, 1, 2, 3, 4, … n.

These numbers play a prominent role in describing confined standing waves. In music they are called the "harmonics" of a sound generator: e.g., the 1st (or fundamental) harmonic, and the 2nd, 3rd , 4th ... harmonics of a violin string. In chemistry, they are called the "primary quantum number" of an atomic electron: e.g., the 1s (or ground) state, and the 2s, 3s, 4s ... 'excited' states of an electron in a hydrogen atom.

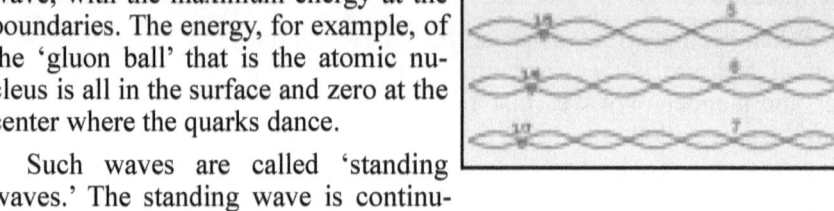

A similar discussion holds for waves that are open at the ends, the cosine wave, with the maximum energy at the boundaries. The energy, for example, of the 'gluon ball' that is the atomic nucleus is all in the surface and zero at the center where the quarks dance.

Such waves are called 'standing waves.' The standing wave is continuous circular motion confined within local bounds, unlike the traveling wave which is globally unbounded. The Greek letter ψ [sigh] is often used to symbolize a standing wave.

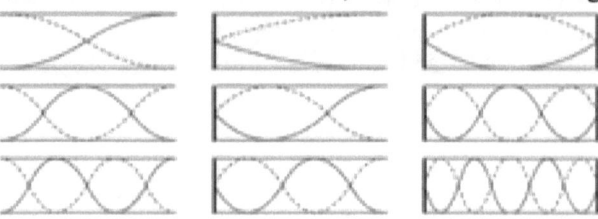

The waves so far have been either open or closed standing waves. Depending on the boundary, we can also have mixed waves, one with a node at one end and an antinode at the other end as in many organ pipes.

Paired waves

There is just one more important property of waves we need to establish. We have just finished a discussion based on flipping the sine and cosine waves about the vertical axis. This next property involves flipping the waves about the horizontal axis.

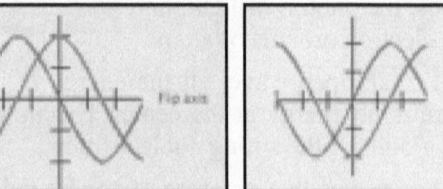

Now both functions change shape. The sine wave just flips into its mirror image as before, while the cosine function flips upside down.

This is a neat illustration of the adage that: "for every up there is a down." If a wave fits waving one way, it will also fit by waving the other way.

"Waves that fit" come in pairs waving in complementary directions. Other than this, the paired waves are identical.

This property of waves plays a central role in the workings of the logos. For every wave that fits, there are two complementary forms:

Where the one goes out, the other goes in; where the one is concave, the other is convex. All things that are given their form by waves come in complementary pairs that fit together.

We will now spend much time describing how all things—animal, vegetable, or mineral—are given form by waves, admittedly much more sophisticated ones than those just discussed. So it will take a while before we are able to properly describe how this wave property relates to the male and female form and its delightful way of fitting together.

For now, we will just note that the great difference between the intuitive behavior of the electromagnetic waves that structure the atom and the non-intuitive chromodynamic waves that structure the atomic nucleus is due to the first being a closed wave with a zero energy boundary (sine waves), while the second is an open wave with a high energy boundary (cosine waves).

Interference

So far we have discussed one wave that fit inside the resonator with an integer number of wavenumbers. When two or more waves occupy the same cavity they "interfere" with each other. This can be either positive or negative.

Two identical waves that fit simply add together at every point—the wave has twice the amplitude. The waves are said to be "in phase." While the amplitude of the wave doubles, the intensity of the wave is quadrupled; if the amplitude is tripled, the intensity is nine times as great, and so on. This is positive, constructive interference of the two waves and is energetically favorable.

Wave amplitude:	1	2	3	4...
Wave intensity:	1	4	9	16...

Negative, destructive interference of waves occurs when the 'up-ness' of one is cancelled by the down-ness of the other. Perfect cancellation occurs when a sine wave is added to the negative sine wave. The resulting wave is zero everywhere, it is no wave at all. The two waves have annihilated each other. This is destructive, negative interference. The two waves are totally "out of phase."

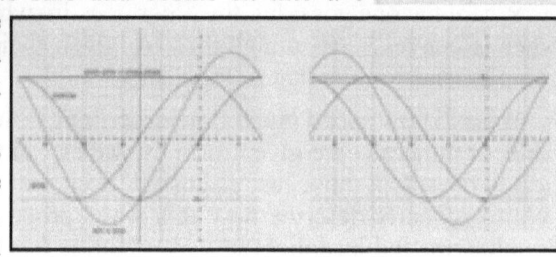

Adding sine and cosine waves of the same period together generates a new curve with a slightly greater amplitude than either. This wave also has a period that is larger, where the complete sine and cosine fit into a θ space, the composite wave has only completed ≈1⅓ wavelengths as can be seen from the diagram. The composite wave is out of phase with both the sine and cosine waves. While waves that are in phase interfere constructively, waves that are out of phase interfere destructively.

Consider a sine wave starting off in a resonator which is 99/100th its wavelength, a non-integer number. The reflected wave is still the opposite of the incoming wave, but now the wave is not at zero when it hits the boundary: the reflected wave is now not the same, it is a different looking wave that is "out of phase" by a small amount with the incoming wave. Things get worse when the reflected wave is reflected back again, the wave is out of phase with the original. The wave is diminished rather than being strengthened.

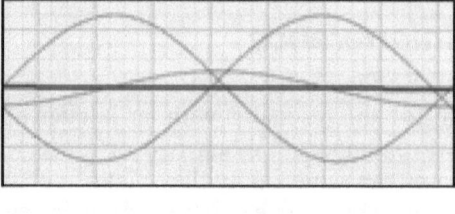

With constant reflection the waves are all out of phase and the result is zero—a wave that does not fit quickly dies away, the resonator will not resonate at that frequency. This is why all standing waves are waves that fit by ½ periods into their bounds.

Harmonious resonance

Next we are going to consider a sound generator that is creating a large set of sine waves of randomly varying frequency—noise as we call it. This is placed as input to a reflecting confined space which has an outlet.

Only those waves that fit within the confines as integer wavenumbers will survive, all of the energy of the random input ends up in the 1st, 2nd, 3rd etc. harmonics. The mix of harmonics in this 'standing chord' is different for different resonators, and is what makes a 'middle C' generated by an organ, a flute, a violin and two human singers sound so different and distinctive.

If the confining space is changed, the standing chord shifts to fit. The time it will take for this sorting out of waves will depend on the speed of the wave. For sound and light with finite speed, this change in the standing wave, the wave that fits, takes a finite time. As we shall later explore, changes in the 'state of the wavefunction' occur within a "pixel of time" no matter the spatial separation involved. As far as current technology is concerned, the change in the wavefunction-that-fits is instantaneous.

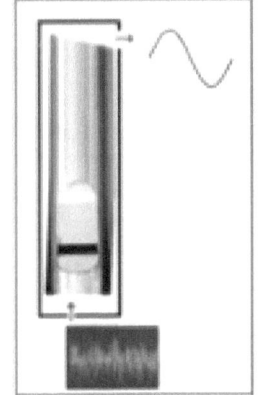

When two waves with wavelengths related by a simple fraction to each other combine many interesting phenomena occur.

Some very interesting patterns of constructive and destructive interference patterns occur even when the traveling waves are the same frequency but coming from slightly different sources, like passing through two slits.

This setup is called a "slit experiment" and it was used to discover the wave aspect of photons (a boson) and electrons (a fermion).

The following diagram shows two waves whose wavelengths are 5/6th different in length traveling separately and then together. They make a beautiful pattern of wave-within-wave. In the case of sound waves, this beautiful pattern is perceived as a melodious chord, a pleasant combination of two notes.

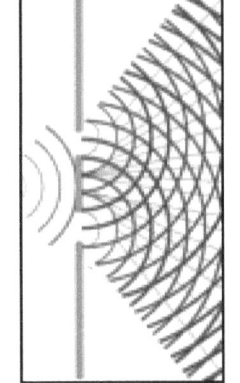

Only the small fractions are perceived as harmonious. The large fractions, such as 10/11 or 11/12 are dissonant and are perceived as unpleasant to most listeners. The small fractions relating wavelengths are considered harmonic and are given their own names in musical theory.

A "chord" in music is when two or more harmonious notes are sounded together. A 'major chord' involves a combination of three harmonious waves. So that all musical instruments can be in tune with each other, there are standard wavelengths, standard frequencies, each with its own name.

A symphony

In music, very intricate harmonious waves

are created. A sophisticated example is a choral symphony being performed in a concert hall. This system involves a great number of subsystems.

There are a relatively small number of 'wave generator' subsystems, the musical instruments and choir members who create the complex sound waves. There are also a relatively great number of 'wave resonators,' the air molecules whose history and form reflect that of the combined waves projected by the generator subsystems.

SYSTEM	INTERACTING SUBSYSTEMS	
	WAVE GENERA-TOR	WAVE RESONATOR
CHARACTERISTICS	FEW MASSIVE IMMOBILE	MANY SLIGHT MOBILE
SYMPHONY	INSTRUMENTS	AIR MOLECULES

The generator subsystems are few, massive and barely move. The resonator subsystems are many, lightweight and mobile. It is the resonator subsystems that 'couple' the audience to the symphony performance.

We will find this differentiation between the two roles of subsystems in a system most useful when dealing with the sophisticated waveforms we regularly encounter in everyday life.

COMPLEX NUMBERS

While the sine and cosine functions are similar, they involve looking sideways at circular motion along different dimensions: the cosine is looking along the length side while the sine function is looking along the height side. The only point they have in common is the zero point. They involve different dimensions, and this difference must captured in the mathematical vocabulary.

The description of this difference involves the concept of operators. We have already seen that in multiplication there are two different roles, one the subject that is acting on the object. The technical way to express this is that an 'operator' is acting on an 'argument.' In the self-acting process of squaring, the number takes on both roles.

The real number line of the cosine axis is rotated anticlockwise by 90° to be the real number line of the sine axis.

The Rotation Operator

This operation "rotated by 90° from the real axis" is indicated by multiplying by letter, 'i.' This is the "rotation operator" that rotates anything it multiplies by 90°. The operator '–i' rotates everything by 90° in the opposite direction.

Rotating by 90° and then again by 90° in the same direction, i2, ends up back on the real line, but in the opposite direction. This is -1, the negative operator that flips everything by 180°. We get to the same place if we do this in the opposite direction. We have

$$-1 = +i2 = -i2$$

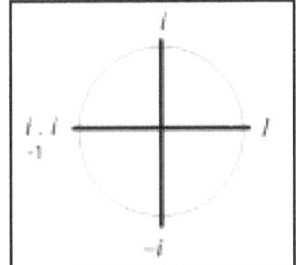

which is why 'I' is also called the 'square root of minus one.'

Performing a third 90° rotation brings us to '–i' at ¾ around the circle, and this is just '–i'.

A final and fourth 90° rotation brings us back to the starting point at +1. This is a cyclic group of 4 rotational operators related in the following way.

$$i1 = i \qquad i2 = -1 \qquad i3 = -i \qquad i4 = +1 \; : \quad i5 = i$$

This is the group of orthogonal rotation operators:

+1   the identity operator
+i   the 90° anticlockwise rotation operator
–1   the 180° flip operator
–i   the 90° clockwise rotation operator

The sine wave occupies an orthogonal dimension to the start axis of the cosine wave, and this distinction result is symbolized as:

$$\cos x + i \sin x$$

In adding the two real numbers, the + signifies bringing the two numbers together and uniting them while it is assumed that both numbers are positive—the form x + y is just a simplification of (+x) + (+y) that conflates adding with direction.

This brings us to a discussion of one more type of number to add to our menagerie of concepts that emerge inevitably from the concept of Absolutely Nothing; the integers, rational, irrational and transcendental numbers of linear extension and angular rotation. These are the 'complex' numbers that combine both linear and angular motion into a single number.

Negative numbers

The number −1 is not so mysterious in terms of addition and subtraction, the number line just extends in the opposite direction to the counting integers. It is, however, when considered in the subject role of multiplication. On the real number line, the object number first shrinks to zero and then extends outwards from the other side. This works for even a huge number such as a google $-1 \times 10^{100} = -10^{100}$. As zero plays a unique role in multiplication, this is hard to comprehend.

But when considered as a rotation operator of 180° (π radians, ½ period), multiplying by −1 just rotates the number around and this problem of going to zero does not arise.

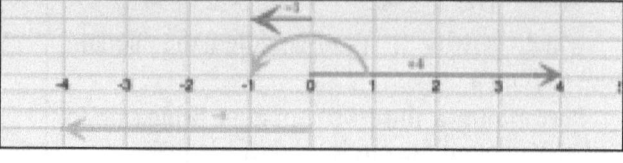

A number rotated by 180° has a zero rotation when rotated by another 180°. This is a much more intuitive understanding of the 'rule of signs' than the mnemonic, "Minus times minus is a plus, for reasons we will not discuss."

Note that this flip back and forth is indifferent to clockwise and anticlockwise direction (or mixtures of the two). It is only rotation by ½ that is sensitive to direction, and ½ plays a special role in the properties of waves.

Imaginary numbers

With the concept of numbers rotated by 180°, it is a simple step to numbers rotated by 90°, numbers that are on a number line that is orthogonal to the real line, and intersecting the real axis at zero. Multiplying the number 4 by i results in the number 4i.

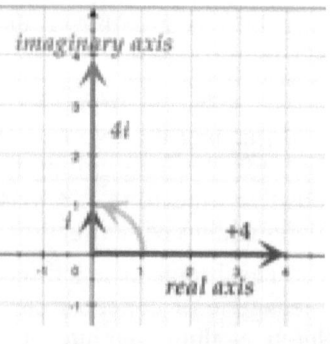

Such numbers are called 'imaginary numbers' (the name again suggesting the struggles the pioneers had in accepting such numbers). Multiplying a number rotated by 90° by a number also rotated by 90° ends up with a number rotated by 180°, a negative number, so the square roots of negative real numbers are not found on the real number line, they are imaginary numbers.

The real and imaginary numbers only have one point in common, the integer 0. Rotation by 90° is sensitive to clockwise and anticlockwise. Going in the clockwise direction is the operator −i. Doing this twice also ends up on -1, so −i is as much the square root of −1 as is +i.

$$+i \times +i = -1 \quad \& \quad -i \times -i = -1 \quad so \quad \sqrt{-1} = \pm i$$

Complex numbers

It is then but a small step to numbers with any linear size, m, (its magnitude) and any rotation, a, (its amplitude). These are the complex numbers, and they not ordered points along a linear line but points on the 'complex plane'. It can be indicated by an arrow of length m with its origin at zero, making an angle a with the real positive axis, the zero for rotation. This is the 'polar' form for a complex number, (m,a).

Complex numbers can also be resolved into real and imaginary numbers components, the cosine and sine components times a scaling factor. This is the rectangular or Cartesian form of a complex number, (x+yi)=n(cos a+isin a) with the n as a scaling factor.

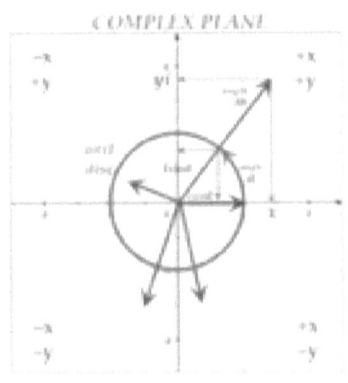

The polar and rectangular forms of a complex number, z, are related by this set of relations (where the arctan function is "the angle whose tangent—the ratio of sine to cosine—has the value"):

The great mathematician Leonhard Euler used the infinite expression for 'e' recently mentioned to explore what happened when 'i' appeared in the exponent position. By regrouping, he ended up with the infinite expansions for the cosine and imaginary-sine functions. He established 'Euler's Formula when the angle, a, is in radians. The angle of ½ period, π radians, then gave him 'Euler's Identity' (which many think the most beautiful formula in math) linking the transcendental numbers 'e' and π to the integers.

$$z = m @ a \qquad = m(cis\ a)$$
$$= x + yi \qquad = m(\cos a + i\,si)$$
$$m = \sqrt{x^2 + y^2}$$
$$a = \arctan \frac{y}{x}$$

This leads to yet another way of expressing a complex number as an imaginary power of that remarkable transcendental number, e. This gives (at least) five different ways of expressing the same complex number, z.

$$e^{ia} = \cos a + i \sin a$$
$$e^{i\pi} = -1$$

Each expression has its advantages and disadvantages, but is it always possible to move easily between expressions using the identities already discussed.

For instance, adding and subtracting complex numbers is very simple in the rectangular form of x and y, while multiplication and division are not. Conversely, it is simple to multiply and divide complex numbers in the polar form of m and a, while addition and subtraction are not. As the not-simple expressions are rather ugly, I have not included them in the table.

$$z = me^{ia}$$
$$= m@a$$
$$= m(\cos a + i \sin a)$$
$$= m\,\text{cis}\,a$$
$$= x + yi$$

The complex numbers are 'complete' in that nothing you can do to them results in something that is no longer a complex number. The real numbers are not complete, for the square roots of all the negative numbers are not to be found on the real line. Complex numbers are found throughout the math level of the Abstract Realm,   and they are extensively used in the modern scientific explanation of the workings of the Physical Realm.

	rectangular	polar
$z_1 + z_2$	$(x_1 + x_2) + i(y_1 + y_2)$	
$2z$	$2x + i2y$	
$z_1 - z_2$	$(x_1 - x_2) + i(y_1 - y_2)$	
$z_1 \times z_2$		$m_1 \times m_2 @ a_1 +$
z^2		$m^2 @ 2a$
z_1/z_2		$m_1/m_2 @ a_1 -$

The positive real numbers are 'honorary' complex numbers with zero rotation, so the complex numbers inherit all the emergent properties of the linear real numbers. The combination of linear and angular, however, gives the complex numbers a host of new emergent properties. One example is the Mandelbrot set of complex numbers.

Mandelbrot set

$$z_{n+1} \xleftarrow{\;n \to \infty\;} z_n^2 + z_0$$

Creating the Mandelbrot set involves iteration of squaring and adding a complex number, z0, such as in the example shown, and seeing what happens with the result of the repetitions. If the result is bounded, the number, z0, is in the Mandelbrot Set. If the result zooms off to infinity in any direction, the number, z0, is not in the set.

In this illustration of the Mandelbrot Set (MS) of complex numbers on the complex plane, the members of the set are colored black, while the numbers not in the set are shaded according to how rapidly they zoom off to infinity.

The set is symmetrical along the imaginary axis (as befits the equivalence of clockwise and anticlockwise) but is not on the real axis which goes roughly from −1.8 to +0.25.

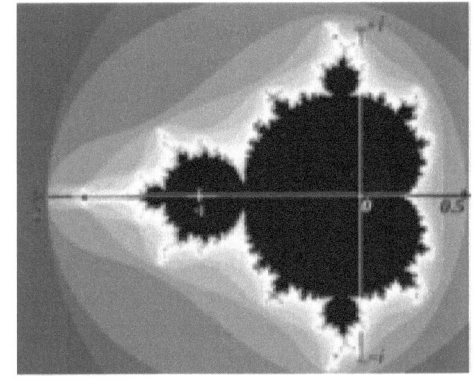

The fascination of the MS is the fractal, irregular boundary of the set. 'Fractal' is the property of having a structure on every level of magnification, while irregular is the property of never quite repeating the pattern.

Numbers that differ by billionths, or trillionths, etc., can be segregated by the boundary. In the illustration, a spot on the upper left of the MS is magnified so that the final one is a 100,000,000 times magnification of the first.

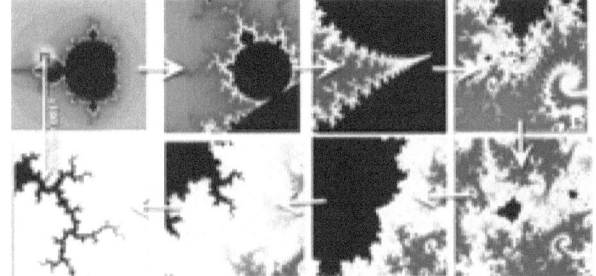

Harmonic Primes

Another of the fascinating properties of the complex numbers was discovered when they were combined with the 'harmonic series.'

This is the name that is given to the sequence of numbers created by adding up the reciprocals of the integers. While the rate of increase is slow, it was proved that the series was not bounded, you could make the sum any size you pleased by taking the sequence out far enough.

$\frac{1}{1}$	1
$\frac{1}{1} + \frac{1}{2}$	1.5
$\frac{1}{1} + \frac{1}{2} + \frac{1}{3}$	1.8\overline{3}
$\frac{1}{1} + \frac{1}{2} + \frac{1}{3} + \frac{1}{4}$	2.08\overline{3}
$\frac{1}{1} + \frac{1}{2} + \frac{1}{3} + \frac{1}{4} + \frac{1}{5}$	2.28\overline{3}
$\frac{1}{1} + \frac{1}{2} + \frac{1}{3} + \frac{1}{4} + \frac{1}{5} + \frac{1}{6}$	2.45
$\frac{1}{1} + \frac{1}{2} + \frac{1}{3} + \frac{1}{4} + \frac{1}{5} + \frac{1}{6} + \frac{1}{7}...$	2.59\overline{285714}

$$\sum_{1}^{n} \frac{1}{n} \xrightarrow{n \to \infty} \infty$$

It was also known that the reciprocals of the integers-squared created a bounded series, and the genius of Euler was to show that this limit involved π-squared as noted above. Nowadays this limit is called the zeta of the variable, s, so Euler's result is the "zeta of 2" and infinity is the zeta of

1. His method allowed the calculation of the zetas for all the positive even integers. For the odd integers, this method fails, however, and the only thing known is that they have an (unknown) limit.

$\frac{1}{1^2}$	1
$\frac{1}{1^2}+\frac{1}{2^2}$	1.25
$\frac{1}{1^2}+\frac{1}{2^2}+\frac{1}{3^2}$	1.36$\bar{1}$
$\frac{1}{1^2}+\frac{1}{2^2}+\frac{1}{3^2}+\frac{1}{4^2}$	1.4236$\bar{1}$
$\frac{1}{1^2}+\frac{1}{2^2}+\frac{1}{3^2}+\frac{1}{4^2}+\frac{1}{5^2}$	1.4636$\bar{1}$
$\frac{1}{1^2}+\frac{1}{2^2}+\frac{1}{3^2}+\frac{1}{4^2}+\frac{1}{5^2}+\frac{1}{6^2}$	1.4913$\bar{8}$
$\frac{1}{1^2}+\frac{1}{2^2}+\frac{1}{3^2}+\frac{1}{4^2}+\frac{1}{5^2}+\frac{1}{6^2}+\frac{1}{7^2}...$	1.51179...

$$\zeta(2) = \sum_1^n \frac{1}{n^2} \xrightarrow{n \to \infty} \frac{\pi^2}{6}$$

$$\zeta(s) = \sum_1^n \frac{1}{n^s} \xrightarrow{n \to \infty}$$

This 'sum of reciprocals' formula for calculating the zeta only works for real numbers greater than 1, but another method for calculating the zetas, established by Bernhard Riemann, allowed for the calculation of the zeta for any number, including the complex, and every number except the integer 1 was found to have a finite zeta (that might be a complex number).

$$\zeta(-2) = \zeta(-4)... = 0 \qquad \text{trivial zeros}$$

$$\zeta(\tfrac{1}{2} \pm iy) = 0 \qquad \textit{zeta zeros}$$

There were certain numbers, that when plugged into this formula, everything cancelled out and the result was zero. These numbers are called the 'zeros of the Riemann zeta function'.

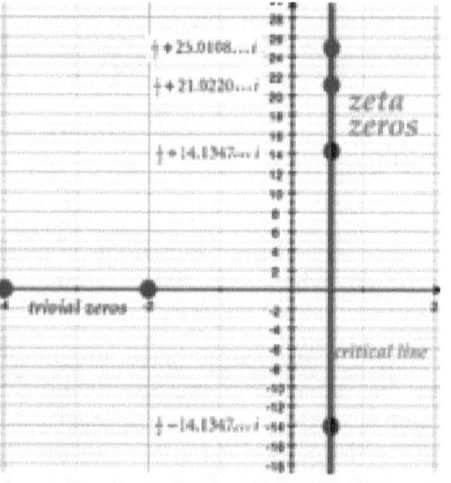

While the positive even integers have a 'closed form' for their zeta involving powers of π, the negative even integers are all zeros of the zeta function. This infinite set of negative even integers is called the 'trivial zeros.'

The non-trivial zeros, and is what is invariably meant by the 'zeta zeros,' is the infinite set of complex numbers that also give a zero when plugged into Riemann's extended zeta function. He proved that if x+iy was a zero, then so was x–iy, and his as-yet-unproved Riemann Hypothesis was that all infinity of these complex numbers had the form, ½±iy. While the trivial zeros are regularly spaced out on the negative real axis, the zeta zeros are irregularly spaced out on the x=+½ imaginary axis, called the 'critical line.' As the size of the imaginary component (ignoring signs) gets larger, the number of zeta zeros along this 'critical line' increase, the zeta zeros get more numerous. This is the opposite behavior to the irregular spacing of the primes which become less numerous as the size of the integers gets larger.

All this brings us back to the integers and the primes, for this was what Riemann was actually exploring, the error function in the prime counting theorem. What he discovered was that the increasing spacing of the primes along the real axis was intimately connected to the decreasing spacing of the complex zeta zeros along the imaginary axis. That the 'error function' in the Prime Number Theorem—which links the distribution of the primes to the transcendental number 'e' in its logarithmic form—was a function of the zeta zeros. The more zeta zeros included, the more accurately was PNT adjusted.

$$\pi(n) \sim \frac{n}{\log n} \qquad \text{the PNT}$$

$$= \frac{n}{\log n} + \text{error}$$

$$= \frac{n}{\log n} + f(\textit{zeta zeros})$$

We have already seen how jagged waves, such as the sawtooth, can emerge from a combination of smooth sine waves. In a similar way, the irregular, jagged graph of π(n) emerges as the contributions of more and more zeta zeros are added to the PNT.3 In the limit, the prime counting function, π(n), emerges.

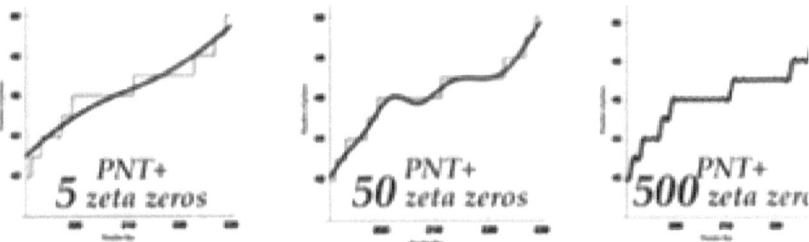

Even deeper and sophisticated mathematical connections between the primes occur beyond the simple math I am using, so you will have to read his book to find out what prompted an author to exclaim in puzzlement: "What on earth does the distribution of prime numbers have to do with the behavior of subatomic particles?"4

ABSTRACT HIERARCHY

This concludes our brief look at the mathematical levels at the foundations of the Abstract Realm. Some thinkers do not accept that the Abstract Realm as an objective aspect of reality, and consider Mathematics as a

---

3 Illustration from: Rockmore, Dan (2007). Stalking the Riemann Hypothesis: The Quest to Find the Hidden Law of Prime Numbers (p. 88). Vintage.

4 Derbyshire, John (2003-04-15). Prime Obsession (Kindle Locations 5365-5366). Joseph Henry Press. Kindle Edition.

creation of human beings while the Physical Realm is all there is to objective reality.

This is a rather provincial view, however, as human beings have only been around for <100,000 years, while the Universe has been around for ~13.5 billion years, according to current science. Eugene Wigner called it "the unreasonable efficacy of mathematics in the natural sciences." He said this because all of what are called the 'hard' sciences express their concepts in the language of mathematics. (Sciences, such as biology, that rely on fuzzy 'natural' language, such as English, to express their concepts aspire to hardness, but are in the meantime considered 'soft.') Furthermore, all scientists firmly believe, quite rightly, that the mathematical laws uncovered in this age are exactly the same as those at work throughout the billion-year history of the universe when no humans were around. The universe runs by mathematical concepts that preceded humans, and to think otherwise goes against the facts.

An attempt to read an advanced paper on fundamental physics should be sufficient to convince even a devout materialist that the writers consider their subject to be pure mathematics. Modern physics considers the 'fundamental particles' out of which atoms are constructed to be much more 'bits of pure math' than 'bits of pure matter.' There is, for example, an area of simple mathematics called 'group theory' (dealing with sets of things that transform into each other, but not discussed here) that successfully predicts what comes flying out of high-energy smash-ups in particle accelerators. This can only support the concept that entities in the Physical Realm are actually constructed out entities in the Abstract Realm.

The Physical Realm, we conclude, is made of math and run by math. Naturally, such a view precludes any worry about "the unreasonable efficacy of mathematics."

Starting with Absolute Nothing, we have the inevitable appearance of the integers and how they interact with each other (addition and multiplication) and upon themselves (squaring). The interactions at this level along with the concept of rotation leads to the inevitable emergence of higher levels in the hierarchy, the rational, irrational, transcendental, imaginary and complex numbers. Waves are the emergent properties of combinations of linear and angular motion.

On these foundations are the 'higher' mathematics based dealing with very sophisticated entities with many subtle emergent properties.

The basic tool in mathematics is the 'proof'—such as we saw in the proof that the square root of two or any prime number cannot be the ratio of two integers. We can restate as 'a universe in which the square root of two is a rational number is inconceivable.' Using such language, we can state that it is inconceivable that the universe does not have an Abstract Realm that is as real as any other realm. A realm that exists independent of any human exploration of its structure.

Emotional waves

We have seen in the zeta zeros that the hierarchical structure of the Abstract Realm can get very sophisticated. We will now look briefly at just how sophisticated the highest levels might get.

One indication that waves play an important role at very high levels of sophisticated emergent properties is to be found in music, which can convey a wide range of emotions. A punk-rock band pounding out their distaste for society, and a choir & orchestra affirming the joy of life in the climax of Beethoven's 9th Symphony, are equally loud, but the emotions conveyed to the audience are quite different.

Scientists have minutely scrutinized music to see what it is made of. The are by now throughly convinced that all music is pressure waves in air. Whatever an emotion is, it must be conveyed by the shape of the wave, as the 'air' aspect to music is immediately discarded by the listening ear and only the wave is passed on, transformed in turn, into the vibrations of a membrane, the quivering of tiny bones, another membrane, pressure waves in water. This last is so shaped that standing waves form with their antinodes of energy at different places, a 'Fourier Analysis' of the complicated wave at each instant into its component sine waves. Cells lining this shaped water have little flags, called cilia, on their surface. Such cilia located at an antinode wiggle about in the energy there, while cilia located at a node do not as there is no energy.

The amount of energy determines how vigorous the wiggle, call it x, and the cell responds by reporting a digital number up the neural hierarchy that for its set position, call it period a, the energy in the water pressure wave is sin2(a)x. Each patch of cells does this for all frequencies.

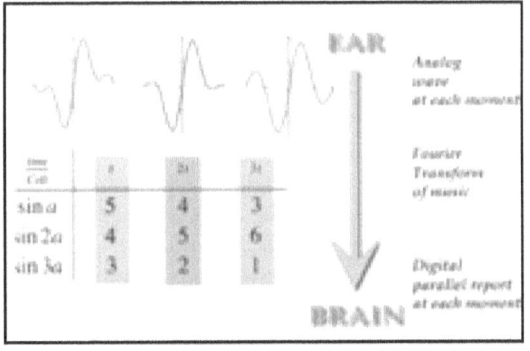

The ear converts the sound wave each moment into the sum of a series of sine-squared waves, and converts this into digital form that is sent, in parallel, up the auditory nerve for further processing. The details of what happens next is still being worked out, but the end result is that we 'hear' the analog wave not the digital numbers, so the wave must be recreated by a reverse transform from digital information to an analog wave.

A simple conclusion from all this that an emotion is the form of an intricately-structured wave. Simply put, an emotion is the property of a wave with a fractal structure akin to that of the Mandelbrot Set.

We can conclude that these sophisticated waveforms in the higher levels of the Abstract Realm have the emergent property we call 'emotion.'

Structure of Man

What about the very highest levels in the Abstract Hierarchy? Can we say anything meaningful about them. Modern science is well in advance of mathematics in answering this question as a result of (metaphorically) taking a human being to bits and seeing what it is made of.

The answer (and we will soon deal with this in detail) turns out to be remarkably simple as scientists found there were only two things:

There found a set of 'fundamental particles,' FP, which, as mentioned, are treated as mathematical entities in modern science. The properties of the FP are measured with real numbers, and divide the FP into two classes that obey a different math: The 'force' particles (bosons), and the 'matter' particles (fermions.)

They found a 'wavefunction' that was strictly determined by a mathematical law and could only be described with complex numbers. The wavefunction determined what happened to the FP of both kinds, but only indirectly. Rather than the wave directly determining what happened to the particles, it was the intensity of the wave that determined what the particles would do over time.

The conclusion was that if the wave had an intricate form then the particles would take up the mathematically-related form of the intensity of the wave. If the wave was a simple sine, the form of the particles over time would be sine-squared.

It is bounded waves that confine particles together long enough to be considered a 'system' and given a name (while the transitory encounters of traveling waves are usually not so dignified). The number of particles of both kinds involved in the make-up of a human being is enormous—trillions of trillions of trillions of them—and many of them are on their way in or on the way out of the human being. In order to keep track of this enormous number of particles we employ a hierarchy of names—such as organ, cell, molecule, atom, etc.—but they are all just various numbers of particles in a hierarchy of waves.

We have already discussed how simple waves readily combine into more complicated waves. On a human timescale, the zillions of particles making up the human body move as if they were confined and organized by a unified wave, rather than as a collection of isolated waves. Watching a ballerina pirouette or an athlete sailing over an obstacle, it is clear that a unified wave is moving all those zillions of particles in a unified manner.

So a human being is a unified wave, described by complex numbers, that confines and structures a zillion or so fundamental particles, described by real numbers. Nothing more, nothing less is there.

We have already established that emotions are complex wave forms. In everyday language we say that it is our mind that experiences emotions, that it is the "I Am" who is experiencing the emotion (usually along with a lot of other stuff going on in the mind as well).

We can tie all this together by equating the the human mind with the unified wave, and the human body with the intensity of this wave as expressed in particles.

The human mind, in this view, is a wave at the pinnacle of possible waves, with the emergent properties of "I Am" capable of intellect, emotion and will.

As this utterly sophisticated hierarchy of waves is also to be found at the pinnacle of the Abstract Realm, we conclude that the emergent properties at this level include 'I Am,' emotion, intellect and will. This is what is usually called God.

So, starting with the concept of Absolutely Nothing it is inconceivable that we do not end up with a God of heart, intellect and will.

We shall assume that this God initiated the creation of the Substantial Realms out of the Abstract Realm, and created an abstract entity to run the creation along a track to the fulfillment of a grand purpose. Questions about the purpose, how things got off the track, and what is being done to get things back on track, are best dealt with in theology. All I will say on this topic, for now, is the question, "If you were an eternal being with the capacity for love, wouldn't you want to spend eternity with those you love?"

We will call the abstract entity created to run the universe, the Logos. It is also hierarchical and, at its lower levels, embraces the mathematical principles that scientists call 'Natural Law.' We will explore what natural law is and its similarity to spacetime and physical systems in Book Two.

The Abstract Realm is founded on, to put it in the simplest mathematical form, the inevitability of Absolutely Nothing leading to the concept of one, and so on. The two substantial realms are founded on a different principle, that zero can be separated into two equal but opposite quantities (which, if allowed to combine would amount to nothing).

Two such possibilities are $0=+1-1$ and $0=+1+i-1-i$, the first creates a line while the second creates a unit complex plane. Using a multiple of this second method, the Logos first separated a zero of the Abstract realm into four orthogonal complex units. Then the Logos asymmetrically twisted apart this set of eight entities into two sets of four; the 's-metric' with one imaginary and three real axes, and the 'p-metric' with one real and three imaginary axes. The asymmetry was such that the s-metric had a twist to the right, R, while the p-metric had a twist to the left, L.

Under the direction of the Logos, these two metrics went their separate ways and expanded into the two Substantial realms, called the Physical realm and the Spiritual realm. In both metrics, there was a wave of period 1 and velocity 1. In the p-metric with the three imaginary axes, the intensity of the wave was −1, while the intensity of the wave in the s-metric with the three real axes was +1. This difference gave the two metrics entirely different histories under the direction of the Logos, with many properties of the two realms having complementary relationships such as +n and −n, or n and 1/n.

As the history of the p-metric is known in some detail, we will spend most of the rest of this work studying the structure and history of the Physical realm. We will only return to the s-metric towards the end of the work as it is currently almost unknown to the sciences. I say 'almost' since one aspect of it has recently emerged into scientific purview as the 'Dark Energy' that is accelerating the expansion of the visible universe.

2.

FUNDAMENTAL PHYSICAL ENTITIES

While the destiny of the p-metric was the substantial Physical realm, its origins were not at all 'substantial' but Abstract. At the moment of creation, the Physical realm was the p-metric with a wave of −1 intensity. The theory of Special Relativity defines a quality in this metric called 'separation' related to the four axis by the Pythagorean relationship.

This speck of −1 is called a the 'false vacuum' in modern physics, and it has what is called a 'negative pressure' in the theory of General Relativity.

$$\text{metric} \quad \pm 1_t \pm i_x \pm i_y \pm i_z$$

$$\text{separation} \quad s = \sqrt{\left(\pm 1_t\right)^2 + \left(\pm i_x\right)^2 + \left(\pm i_y\right)^2 + \left(\pm i_z\right)^2}$$

$$= \sqrt{1_t - 1_x - 1_y - 1_z}$$

This negative pressure caused the p-metric to enter what is called "a period of inflation" in which the separation increased exponentially to very large values. This much enlarged 'four dimensional' construct then entered a 'period of braking' when much of the rate of expansion was converted into a set of wrinkles and twists in the four dimensions, called 'fundamental particles,' in a period called the hot Big Bang.

The 'fundamental particles' of modern physics are more accurately called the 'fundamental entities' of the substantial physical realm, for while they have many aspects of a particle (such as location) they have at least as many wave aspects to them that are not local.

No matter the complexity, size and sophistication of the systems studied in the Physical realm as it currently understood by current science—be they galaxies, stars, planets or dust; be they bacteria, plants, animals or human—they are all composed of just a few kinds of fundamental entities interacting by exchanging a few other kinds of fundamental entities.

In order to understand things, we need to understand the things they are composed of, and how these things come to be and how they behave. For this reason we will have to spend a lot of time discussing the fundamental insights of modern science into what matter is composed of.

***

First we will discuss the nature of the fundamental entities and their relationship to the Logos (natural law) and only then return to what caused the braking of the global expansion into local twists and wrinkles of space-time in the hot Big Bang.

SPACETIME

The impulse of the global inflation was braked and turned into local wave-like deformations of the p-metric with an intensity, measured by real numbers, called the 'external energy' of these fundamental entities. They also retained an 'internal wavefunction' aspect, measured with complex numbers, that retained an imprint of the global wave of the inflationary period called 'entanglement'. The external energy is equivalent to what current science calls the 'particle' aspect of the fundamental entities, and the internal wavefunction gives rise to the 'wave' aspect.

To summarize the rest of the chapter which will deal with both aspects in detail: The wave tells the particle what to do; the interactions of the particle tell the wave how to change.

Classical science does not have the concept of an internal causal wave and focused solely on the external interactions. It was only with reluctance that experiment after experiment necessitated its inclusion during the gestation of current 'quantum' science. "In a sense, the difference between classical and quantum mechanics can be seen to be due to the fact that classical mechanics took too superficial a view of the world: it dealt with appearances. However, quantum mechanics accepts that appearances are the manifestation of a deeper structure ... and that all calculations must be carried out on this substructure."[5]

We will discuss the external energy first, as it has been explored by science for many centuries, and then discuss the internal wave aspect which has only been explored for less than a century.

The external energy of a fundamental entity is tied up in the two types of p-metric deformation called 'boson' and 'fermion'. Colloquially, these are sometimes called the relatively evanescent 'bits of force' and the relatively immutable 'bits of matter.' One remarkable finding of recent physics is that the all the intricate complexity of the everyday world in the Physical realm is composed of just these two types of fundamental entities, albeit both in enormous numbers of them. It is as if we had discovered that the world was constructed out of just two types of tiny Lego building bricks.

To understand the difference between the external aspect of these two types of entities we need a little mathematics from an area called 'topology.'

Topology of spacetime

As described, the p-metric has a Left-handed structure involving one real dimension, labelled t, and three imaginary dimensions, labeled x, y and z. While the plus and minus directions along the 'spatial' imaginary axis

---

5 P. W. Atkins, Quanta (2nd ed.), Oxford University Press, Oxford (1991), p. 348.

are very similar, as illustrated by the similar behavior of +i and –i, while the plus and minus directions along the real 'time' axis are not similar, as illustrated by the differences between +1 and –1. At the moment of creation, the wave had a velocity of 1 in the spatial dimensions and a velocity of 0 in the t dimension, as did the local deformations created during the braking period.

The x, y and z spatial dimensions get twisted up with each other during the chaos of the braking period, and it is the two basic ways that dimensions can be topologically intermixed by twisting, generating the two types of fundamental entities.

The external particle aspect of a fundamental entity involves distinct 'twists' in the spatial dimensions along the time dimension. The two possible types of twists correspond to the two types of particles: A boson involves one or more "oriented" twists, while a fermion involves one or more "non-oriented' twists.

To simplify, we can approach this impossible-to-illustrate four-dimensional situation with a familiar three-dimensional example.

Consider a transparent plastic band with width along the ±x-axis and length along the ±y-axis, but with a zero extension in the z-axis. The x-y plane has two distinct sides to it, one facing in the +z direction while the other is facing the –z direction. (In the diagram, these two are shaded blue and pink solely so we can keep track of them.)

We now perform these three operations on this strip:

Cut the strip across the x-axis, separating out a top edge and a bottom edge along the x axis.

Rotate the bottom-x-axis in the z-z plane around the y-axis. This rotation can be in in either of two directions, +x to +z, or +x to –z. The direction illustrated is +x to +z and is designated a Right-up rotation about the y-axis. Rotation of the x-edge in the reverse direction about y is designated a Left-down rotation.

There are just two ways of aligning the rotated split x-axis together so that the third step can take place:

The bottom-x-axis makes a 180° turn about the y-axis in the z-axis.

The bottom-x-axis makes a 360° turn about the y-axis in the z-axis.

The top- and bottom-x-axes are sealed back together into a single x-axis.

The strip has now got a topological twist in it that can only be removed by performing the three operations in reverse order (and the reverse of cutting is sealing. These topological operations involve three orthogonal directions. In this simple example, the three directions involved are just the three spatial dimensions, but more complicated situations are possible:

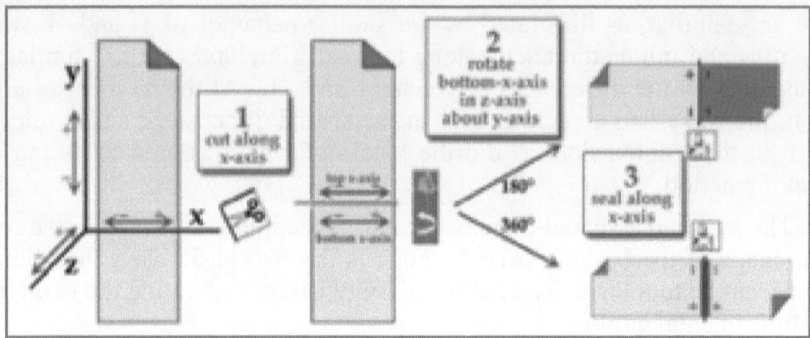

Cutting and sealing along the x-axis

Rotation about the y-axis

Rotation in the z-axis

This topological operation in this simple example alters the three spatial dimensions in the following ways:

Z-axis: This dimension of the strip is unaltered, it is just the same after the operation as it was before.

X-axis: The x-dimension has either:

a discontinuity of + and − directions along it, or

a wave-like patch of disturbance along it.

Y-axis: The effect on a directed x-segment (an arrow in the illustration) making a circular motion around the y-dimension in the z-dimension is either:

Flips the arrow by 180° into its opposite direction. This is called a 'non-oriented' y-axis, and the two-sided strip in the example has become a non-oriented, single-sided Moebius Strip. We can say that the y-axis has become 'disoriented' by the topological operation and has a 'spin' of ½. The spin can be Right or Left depending on the way the the x-axis was

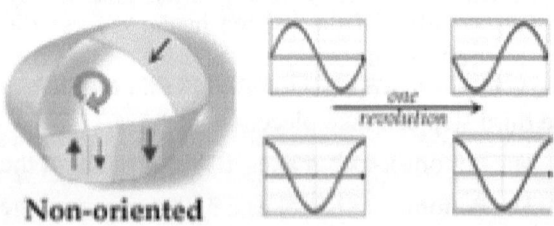

Non-oriented

given its ½-twist. An asymmetrical sine wave, for example, traveling along the disordered y-axis would be altered to a minus-sine wave. A symmetrical cosine wave, on the other hand, would be unaffected by the revolution.

Does not flip the arrow, it just rotates it by a full 360° back to its original orientation. This is called an 'oriented' y-axis. Adding a second ½-twist to the Moebius strip (in the same direction as the first one, because a ½ twist in the opposite direction would just restore all three dimensions back to the original undisturbed state) results in a two-sided oriented band with a wave in it. While a (co)sine wave traveling along the y-axis will be rotated, it will be the same before and after. The oriented axis has a spin of 1, and this can be Right or Left.

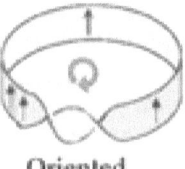

Oriented

Topology of the metric

This simple introduction to Topology will suffice for an overview of the fundamental entities in the substantial entities in the Physical realm are constructed of.

In the simple example in a 3-D space of real dimensions used above, the 'sidedness' of the real dimensions (the x-y plane) and the difference between Left and Right is dependent on the remaining real z dimension. For instance, while it is possible to create a non-oriented single-sided 2-D surface (illustrated) in 3-D space, the surface has to intersect itself (and is called a Klein bottle) and do not come in Right and Left forms.

This limitation is not the case for topological disturbances in the physical metric which, as described, is composed of components of four complex dimensions. A real (or imaginary) component of a complex axis has an inher-

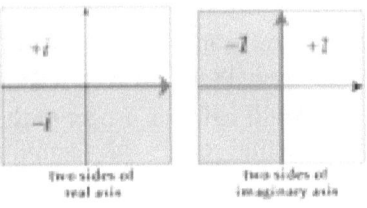

two sides of
real axis

two sides of
imaginary axis

ent 'sidedness' (shown in pink and blue) that is provided by its orthogonal imaginary (or real) axis.

The external 'particle' aspect of the fundamental entities is a topological disturbance created by a slicing of some or all of the three imaginary spatial dimensions of the p-metric, twisting them in the real time axis of the p-metric, followed by a resealing of the spatial dimensions. There are two topological results of this operation:

A 180° twist-in-time results in a non-oriented topological defect. This has an asymmetrical and closed wave aspect, as exemplified by the sine wave, and these fundamental entities with a ½-spin are the set of fundamental 'fermions' each distinguished by the number of spatial dimensions involved in the slicing and sealing.

A 360° twist-in-time results in an oriented topological defect. This has a symmetrical and closed wave aspect, as exemplified by the cosine wave, and these fundamental entities with a 1-spin are the set of fundamental 'bosons' each distinguished by the number of spatial dimensions involved in the slicing and sealing.

The fundamental entities have an external energy. One source is the amplitude of the wave which, as discussed, has an energy that is given by the intensity of the wave, its self-square. This energy in the disoriented, closed fermion wave is zero at the boundary and center of a full-period wave, while in a ½-period standing wave the energy is at the center and zero at the boundaries.

The energy of the oriented, open boson wave is all at the boundary and center in a full-period wave, while in a ½-period standing wave it is all at the boundaries.

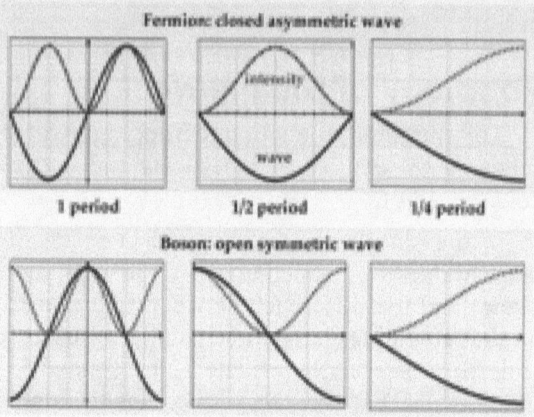

Fermion: closed asymmetric wave

1 period   1/2 period   1/4 period

Boson: open symmetric wave

In discussing sound waves, we mentioned hybrid standing waves in organ pipes with one open and one closed boundary. Both fermion and boson standing waves can fit into such single node boundaries as ¼-period waves.

We will call this aspect to the energy of a fundamental entity its 'amplitude' energy, and, as intensity is the square of the amplitude, a wave of 2sin(a) has four-times the amplitude-energy of a sin(a) wave, all else being equal.

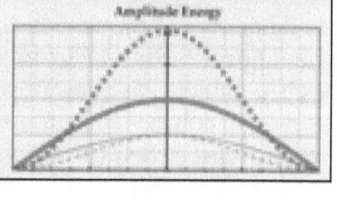

Amplitude Energy

A second source of the energy in fundamental entities is in how quickly the wave is waving, so that a wave that makes two periods while another wave makes only one period has twice the energy, all else being equal. This 'frequency' energy of a fundamental particle is proportional to the fre-

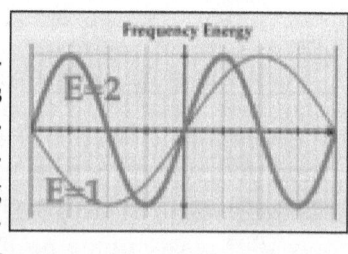

Frequency Energy

E=2

E=1

quency of the wave, so a sin(2a) wave has twice the energy of a sin(a) wave. The energy contributed by the wave is the sum of the amplitude energy and the frequency energy.

A third source of the energy in fundamental entities is in the discontinuities in the topological defect which are of two types:

The first kind of discontinuity occurs in non-orientated topologies. In the simple example we used of the Moebius strip, when sealing the twisted plastic band back together the + and − orientations along the x-axis were irrelevant. We can make our example 'polarity relevant' by making the x-axis a magnet with a North and South end. Sealing the x-axis now involves bringing a North and a South pole together, and this takes energy. The Moebius band now has energy locked up in the sealed x-axis. In an analogous way, segments of the resealed p-metric that have an alignment mismatch also have an energy in them that we can call the 'misalignment' energy. The oriented topologies do not have any misalignment energy.

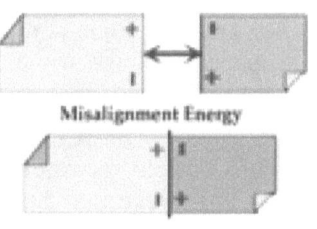

Misalignment Energy

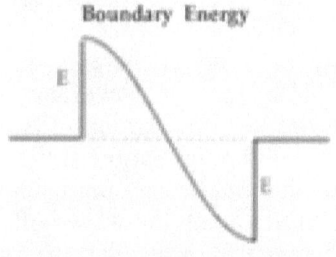

The discontinuity in the oriented topology arises from the open nature of the waves, there is an abrupt discontinuity in the p-metric from waving to not-waving at the boundary of the wave.

To summarize, all the fundamental entities have amplitude and frequency energy. In addition, the fermions have misalignment energy and the bosons have boundary energy. As there are three spatial dimensions in the p-metric in which the external topological twists in time can occur, there are only three types of fermions and three types of bosons, making six types in total.

Boundary Energy

We will just list the six entities here for now, and return to them in more detail after discussing their internal wave aspect.

The names given to the six types of fundamental entities are in no way systematic, and reflect the state of science

D Number	**Bosons** *Oriented, open*	**Fermions** *Non-oriented, closed*
One	Weak (W)	Neutrino (ν)
Two	Photon (γ)	Electron (e)
Three	Gluon (g)	Quark (q)

at the time of their discovery. The customary symbols are also shown: the one-fermion is the Greek 'n' (nu), and the two-boson is the Greek 'g' (gamma); the rest are plain English.

INTERNAL PROBABILITY WAVE

The topological defect (the particle) aspect of fundamental entities involves a deformation of the p-metric, while the wave (the wavefunction) aspect governs just what portion of the p-metric is involved in the defect, where the particle is located. The wavefunction is a complex wave that moves with a constant velocity of 1 through the p-metric.

This complex 'location' wave has a linear size (amplitude) of p, and an angular rotation (phase), of α, together comprising the complex number p@α. The amplitude and phase will vary at locations along the wave, and the intensity of the wavefunction at a particular location, the amplitude-squared, p2, gives the probability of the particle being found at that location.

For example, a particle with a standing wave that has the form of a sine wave will have a standing 'probability density' that has the form of a sine-squared wave.

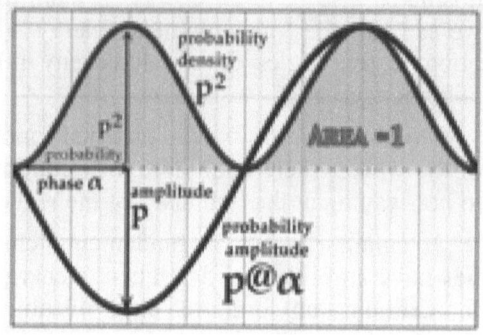

If the particle was always at a single location, the probability of it being there is unity, 100%. In the wavefunction, however, the probability is spread throughout the wave and all that can be said for a single particle in a standing wave is that the probabilities along the wave will all sum to one, it is somewhere to be found in the intensity wave, but we cannot say exactly where. In the realm of calculus (a level in the Abstract Realm we will briefly discuss later) this area under the intensity curve is unity, 100%. For the 1-period sine wave in the example, the technical term is that the 'integral' of the

$$\int_{a=0}^{1} \left(\sin^2 p\right) d\alpha = 1$$

p-axis sine-squared wave over a single period along the α-axis is unity. The wave aspect has a complex-number value everywhere on the p-metric, and this can be zero at locations outside a bounded wave.

The physics of the internal wave reached its apotheosis in the "adding little arrows over-history' methodology perfected by Richard Feynman. (For some reason, he does not think the reader capable of grasping complex number, and calls his little arrows multiplying together 'shrink-and-turn,' shrinking because the amplitude is never greater than one.) This perspective is also called quantum electro-dynamics (QED), the official name for the theory that describes the behavior of electrons and photons in terms of internal probability.

QED is extraordinarily successful and accurate. Feynman has modestly stated that: "The theory of quantum electrodynamics has now lasted more than fifty years and has been tested more and more accurately over a wider and wider range of conditions. At the present time, I can proudly say that there is no significant difference between experiment and theory! ... To give you a feeling for the accuracy [of the quantum description of the electron]: if you were to measure the distance from Los Angeles to New York to this accuracy, it would be exact to the thickness of a human hair. That's how delicately quantum electrodynamics has, in the last fifty years, been checked—both theoretically and experimentally."6

Probability Statistics

It would seem from the above discussion of the probability amplitude, given by the complex number p@α, that only the linear component, p, is an important factor in the probability, p2, and that α is irrelevant. This is true in isolated cases, but when fundamental entities interact with each other (the topic of the next chapter) both have to be taken into account to understand the 'final probability amplitude' and the ending probability of the states after the interaction.

The operations involve the adding and multiplying of complex numbers (sometimes a great many of them) to calculate final complex number and its probability intensity. Complex arithmetic is more sophisticated, and it can successfully predict some quite unexpected results, as illustrated by the Mandelbrot set which is also the result of adding and multiplying complex numbers. (It is, perhaps, being uncomfortable with complex numbers that leads so many writers to speak of the "weird aspects of quantum physics," as if nature is at fault.)

Probability as it is usually understood (or 'regular' probability as we will call it), involves just two simple rules that, as usual, will be illustrated by coin tossing.

A regular coin has a 0.5 chance of landing heads, H, or tails. (This is simply a statement that, if the coin is thrown zillions of times, the probability density will be 50% H and 50% T to whatever accuracy you require).

There are just two simple rules for multiple events in regular probability:

In 'AND' situations, the probabilities are multiplied

In 'OR' situations, the probabilities are added

In either case, the size of the result is the probability.

For example, the probability of throwing a head AND another head, giving HH, is obtained by multiplying ½ times ½ giving the probability of HH to be ¼. (i.e., if a pair of regular coins is thrown a zillion times, the the probability density of HH will be 25%. In the same way, the probability of TT is also 25%.)

The probability of throwing a two heads OR throwing two tails—the 'even' HH & TT throws as compared to the 'odd' throws HT & TH—is obtained by by adding the ¼ probability of each combination together, giving a probability of ½ for an even combination and ½ for an odd.

---

6 Richard P. Feynman, QED: The Strange Theory of Light and Matter, Princeton University Press (1985), p. 7.

Much of the 'weirdness' that so many find in modern physics can be ascribed to the well-established fact that it is the probability amplitudes that are added and multiplied, not the probabilities, which only emerge afterwards in the final squaring step. Otherwise, the rules are almost identical:

There are just two simple rules for multiple events in complex probability:

In 'AND' situations, the probability amplitudes are multiplied

In 'OR' situations, the probabilities amplitudes are added

In either case, the intensity of the result is the probability.

Even though the rules are the same, the switch from the real numbers of regular probability, to complex numbers can produce strange results. For instance, for probabilities greater than zero, adding two of them together to get a zero is impossible. The impossible, however, is possible with complex numbers. Two complex probabilities of the same size but with amplitudes that are 180° out of phase, α and $\alpha+\pi$, $p + p \neq 0$
will add together to create a zero probability.

$$p@\alpha + p@(\alpha + \pi) = 0$$

The complex wavefunctions of bosons combine to create what is called Bose probability, while the fermion wavefunctions combine to create Fermi probability rules; neither of which are as 'regular' probability.

The difference is quite clear in a thought experiment using three types of coins, each of which when tested individually, has a 50% chance of coming up heads. The difference is that when pairs are thrown, one type obeys the rules of regular probability, one type obeys the rules of Bose probability, and one type obeys the rules of Fermi probability.

Put simply, the symmetric waves of bosons add together, the waves combine constructively into one of twice the amplitude and four times the intensity. Bosons have a probability of 1 of entering the same state, so for a boson coin, while the probability of the first H is ½, the probability of the second H is 1.

Two familiar examples of such boson behavior are to be seen in the radio waves emitted and received by TV and radar antennae, and in the laser beams that adjust our eyesight and read our DVDs.

The opposite probability behavior holds for fermion waves. The asymmetric wave flips its sign, so two identical waves attempting to share the same state end up canceling out destructively to zero. This is why the probability of two fermions sharing the same state is zero, it is impossible, it is absolutely forbidden as quantum probability has absolute control over the Physical realm.

The only way two fermions can share the same state is if their waves are opposite, if one is clockwise while the other is anticlockwise. The flip of the asymmetric wave when two identical fermions of opposite spin attempt to share the same state leads to constructive interference into a wave with twice the amplitude and four times the intensity.

As attempting to add a third will inevitably clash with one of the two waves, this reduces to the rule that only two fermions of opposite spin can share the same state. This is called the 'Pauli Exclusion Principle' of natural law, which only allows two electrons of opposite spin in each quantum state. Two somewhat familiar examples of the enforcement power upholding this law can be found at both ends of the size scale: On the tiny level of the atom, this Exclusion Principle is cause of the hierarchical structure to the electrons in atoms, and the chemical interactions they are capable of. On the scale of our sun, all its enormous mass in very old age—in some 50 billion years or so—will be kept expanded in a sphere the size of the earth against the crushing force of gravity (which we will soon discuss) as a cold 'black dwarf' star. The physical universe at 13.5 billion is not yet old enough for there to be any of these to be around, but the transition stage of still-hot 'white dwarf' stars that are slowly cooling off are quite common in our stellar neighborhood.

Fermions have a probability of 1 of entering the opposite state, so for a fermion coin, while the probability of the first H is ½, the probability of a second H is 0 and the probability of T is 1.

Regular probability is based on 'independent assortment' in that the second toss is independent of the first toss. Neither bosons nor fermions show this independence as the second throw depends on the first. This wave-derived behavior

	AND HH (TT)	AND HT (TH)	OR HH /TT	OR HT /TT
Regular	¼	¼	½	½
Bose	½	0	1	0
Fermi	0	½	0	1

only adds to the opinion that quantum behavior is somehow 'weird.'

This table compares the 'probability statistics' for a regular, Bose and Fermi coin.

Note that that the regular probability of a coin composed of zillions of fermions and bosons is just the average of Fermi and Bose probabilities.

Probability Density

The internal wave of both types of fundamental entities can be a bounded wave or an unbounded wave.

For an unbounded traveling wave, the intensity of the internal wave gives the simple probability of finding the particle at a location as the wave travels by.

If an internal wave is separated by some obstacle into two smaller waves, the external particle can appear to be in two places at the 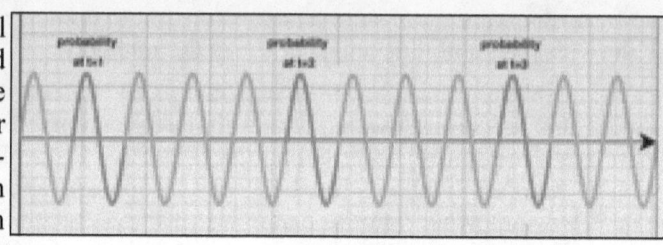 same time as there are now two probabilities moving in tandem, and the particle appears to jump between the two as if the external separation did not exist.

This 'entanglement' of a particle is perhaps the greatest cause of the unease that classical scientists feel towards the 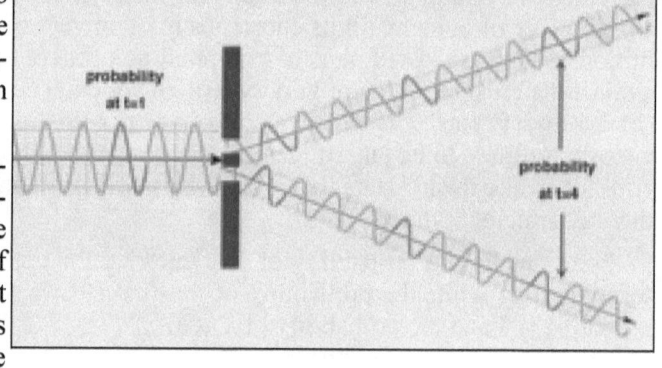 way the universe has been found to function.

The 'development of the wavefunction,' as such changes are called, is determined by natural law, and is the probability. This is the main difference between classical physics and modern physics:

In classical physics, events were thought to be determined by natural law

In modern physics, we know that the probability of events is determined by natural law.

The internal does not, however, determine exactly where a particle is in space at a particular time. This is very different from classical science which thought that the exact location was determined by natural law. It is this lack of external precision that has given rise to the false impression that quantum science is indeterminate.

For standing waves, however, which repeat themselves endlessly, this indeterminism fades away in importance, and the external probability density of a particle is completely determined by the internal wavefunction which, in turn, is completely determined by natural law.

This kind of determinism is called the Law of Large Numbers (LLN). This simply states that the percentage error, or difference, between the probability and the observed probability goes to zero as the number of 'trials' goes to infinity.

A more detailed treatment states that with N trials, the difference between the theoretical probability and what is actually observed almost always falls within the bounds (more technically, within one standard deviation) of plus or minus the square-root of N. So, when throwing a regular coin 100 times, it is quite normal to get 55 heads as this falls well within the bounds of 50±10=40 and 60, and there is no reason to suspect the coin. If 10,000 throws end up with 5,500 heads, however, the coin is most suspect as this falls well outside the boundaries of 10,000±100=9,900 and 10,100.

$$\frac{\sqrt{N}}{N} = \frac{1}{\sqrt{N}} \xrightarrow[N \to \infty]{} 0$$

N	10	100	1,000	1 million	10^{24}
$\sqrt{N}$	~3	10	~32	1,000	1 trillion
$\frac{1}{\sqrt{N}}$	31%	10%	3.1%	0.1%	0.000,000, 000,1%

As the reciprocal gets smaller as the square root gets bigger, the indeterminism can be made as small as required by increasing the number of trials. The fractional difference goes to zero as N gets larger.

For an atomic electron in a bounded standing wave about the atomic nucleus (an orbital) that moves a trillion-trillion times a second (1024 trials), the difference between the intensity of the wave, the probability density, and the actual electron density is essentially zero. It is this aspect of the LLN that leads to such phrases as 'the electron is smeared out in an atom' even though the appropriate probes always find the particle, the topological twist, to be located at a precise point.

Given that it is not usual in nature for the number of trials to be proportional to the time that passes, we can restate this basic principle as: 'Given sufficient time, the form of the internal aspect determines the form of the external aspect.'

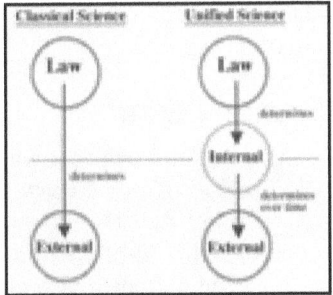

The electron density in an atom determines its physical and chemical attributes. So, on the scale of seconds, it is entirely appropriate to state that the physical and chemical attributes of atoms are determined by the natural law, and indeterminacy drops right out of the picture.

Entanglement

We have already seen in the 'entanglement' set-up of a slit experiment how an external particle can seem to be in two places at the same time. In classical science which does embrace the internal aspect of things, this is, of course, impossible to imagine let alone understand. But in a science in which the internal and external are united, it is the internal that takes precedence and the external just reflects the internal.

A node that has zero extension in the complex dimensions can have a sizable extension in the external spatial dimensions. For a simple bounded sine wave with two 'lobes' separated by a node, the node has essentially a zero extension and the two lobes touch at the central point. If the node has an extension, however, then the two lobes appear separated by a stretch of empty space. This is possible because, as mentioned, the wave in the p-metric is in all the p-metric, and if it just happens to cancel out to zero

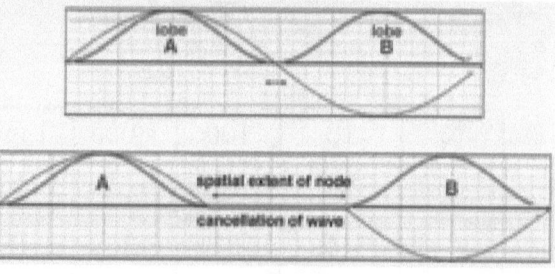

for a stretch and then no longer cancels out at another point, so be it.

The particle aspect ignores any spatial separation, and its probability density continues to fill both lobes as usual. The particle appears to be both locations at the same time. As noted, however, interaction with an appropriate probe always finds the particle to be located at a precise point. The particle aspect of a fundamental entity appears to 'teleport' between location A and location B without ever having any presence in the space separating them. This phenomenon is called 'entanglement.' To make things even 'weirder' to the classical mind, the extent of the spatial entanglement is unbounded, it can stretch across the 13.5 billion light years of the visible universe without any problem, and for entangled particles that are now reaching us that went their separate ways right after the origin of the universe, say 12 billion years ago, the entanglement can stretch 24 billion light years into areas beyond the visible universe from which light has not yet had time to get to us.

Space Travel

As a cultural aside, it was science fiction writers who noted that human colonization of our galaxy, let alone all the distant other ones, would be sluggish and dull if transportation and communication was limited by the speed of light. To make things interesting, they had to invent things such as warp speed, hyperdrives and 'wormholes' through spacetime. The phenomenon of entanglement, when mastered, opens up such possibilities and the rapid colonization of the extraterrestrial planets that are found to be plentiful in our local neighborhood. Nature, as always, provides as there are many natural processes that send entangled particles off at essentially lightspeed in opposite directions, and we are daily showered with a plethora of entangled cosmic ray particles whose other lobe can be many, many light years distant.

As we will discuss in the next chapter, the interaction of a particle alters its wavefunction, and the entanglement is quickly lost once a particle starts interacting with other particles. So the cosmic rays—mainly protons—that penetrate the atmosphere to the earth's surface have lost any of the entanglement they had when entering the solar system. The Moon, however, does not have an atmosphere so it is constantly bombarded by entangled cosmic rays, and we can expect that research into exploiting this natural resource will only be accomplished by a colony on the Moon.

We have already encountered entangled particles on the everyday scale when an electron traveling wave passes through a setup called a 'slit experiment' and it appears to be in two different locations at the same time.

On a tiny scale, we see examples of entanglement in bounded waves. The electrons in an atom are entangled, and the standing waves, the atomic orbitals, can have many nodes in their probability density where the probability is zero. For example, the form of a simple standing wave, such as the 2p orbital, is akin to that of our simple 1-dimensional sine wave where the electron has a zero presence at the center; while the distinctly-odd form of the 3d orbital can only be created

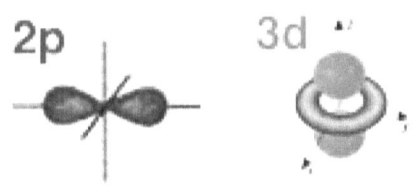

by waves in multi-dimensions. As illustrated by the Mandelbrot set, complex numbers operating on each other are capable of creating a plethora of forms. The electron teleports between the two lobes of the 2p and the three lobes of the 3d as usual, and the probability density of the electron is the intensity of the wave.

We have now looked at entanglement, a property of the internal wave, on three different spatial scales to show that it is a mistake to consider it 'weird':

Nanometer separations in the electron density of atoms

Meter separations of an electron in slit experiments

Gigameter separations of a cosmic ray proton.

Collapse of the wavefunction

Classical science, which insists that an electron has just one location, deals clumsily with the probabilistic aspects of matter by speaking of the 'collapse of the wavefunction.' When the electron moves to a new location and is observed, its extended wavefunction just disappears. This perspective, however, leads into such absurdities as Schrodinger's Cat which is 50% alive and 50% dead before it is observed.

The 'wave is real' perspective, however, has it that the wavefunction changes when the particle interacts, the wave of the particle just alters to a different wave. What happens is that when the particle interacts with the detector, its wavefunction alters and the entanglement is lost.

For example, an electron flipping from a traveling wave in the slit apparatus to a standing wave bound to an atom in a detector. The electron always has a wavefunction, it does not collapse and disappear on observation. This perspective does not lead to absurdities.

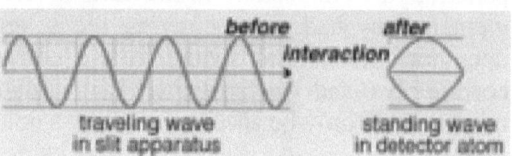

traveling wave in slit apparatus

standing wave in detector atom

before

after

Interaction

INTERNAL AND EXTERNAL

We can now unify the internal and external aspects of the fundamental entities in our listing of them in the physical realm:

Fundamental entities have an 'internal' aspect, a wavefunction that is described by complex numbers. This internal wave is determined by the Logos (natural law).

Fundamental entities have an 'external' aspect, a particle described by real numbers that is a topological defect in space along time. The intensity of the internal wave determines the external probability density of the particle in spacetime.

Discrete Quanta

There is another difference between the math that applies to the internal level and the math that applies to the external particle. We earlier discussed the real number line and the difference between the countable infinity of the rational fractions and the uncountable infinity of the irrational continuum. It is this difference that applies to the internal and external.

	Bosons	Fermions
Internal wave	Oriented, open	Nonoriented, closed
External defect *# dimensions* *One*	Weak (W)	Neutrino (v)
Two	Photon (γ)	Electron (e)
Three	Gluon (g)	Quark (q)

The internal aspect of the p-metric measured by complex numbers is continuous, and it can have rational or irrational values.

The external aspects of the p-metric measured by real numbers is not continuous, and can only have integer rational values. The apparent continuity of the external aspects is actually an artifact of 'resolution.'

For example, a line of one meter that is composed of 1 trillion discrete segments each one a trillionth of a meter long would appear continuous to the naked eye, and it would be impossible to 'resolve' the individual segments. It would, however, be impossible to construct such a line that is exactly the square root of two in length, the best that can be done is a line that is a fragment of a trillionth larger or smaller, but never exactly the length of $\sqrt{2}$.

Color mathematics

Another more everyday example, which will prove most useful a little later on, involves the difference between positive color and negative color or, as it is called in the trade, additive colors and subtractive colors.

On the the screen of computer I am working on, my typing appears as smooth black letters on a pure white background. The colors in the diagram are just as smooth and vivid. When I turn the computer off, however, the screen is black. The smoothness and the colors are all artifacts of resolution.

With a magnifying glass, it will be seen that the computer is actually turning discrete of 'pixels' of Red, Blue and Green on and off. The pixels are so small—72 to the inch—that my eye cannot resolve them individually and I just see the overall impact on the three types of receptors in my eye.

Red, Blue and Green (RGB) are the positive, additive colors. When all the pixels are on, I see a smooth white, and when they are all off, I see a smooth black. When just the Red pixels are on, I see a smooth continuous red. Pixels of two colors add together into a color I also see as smooth and continuous. These are the rules for adding the positive colors to black to get the full gamut of colors we expect from computers and TVs:

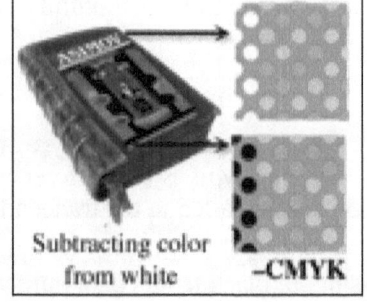

+R+G = yellow   +R+B = magenta   +G+B = cyan

For the subtractive colors, we look to the cover of a paperback book printed on heavy white card. Under a powerful magnifying glass—they will be at least 120 to the inch—the pixels reveal themselves individually to be either Cyan, Magenta, Yellow or Black (CMYK).

The white paper starts off as 100% R, G and B. The RGB colors that affect the eye must now each be reduced, subtracted by an ink, to control the appearance of smooth color at a low resolution.

The ink that absorbs all the R and only lets the G and B from the paper through is a cyan color. The ink that absorbs all the G and only lets the R and B from the paper through is a magenta color. The ink that absorbs all the B and only lets the G and R from the paper through is a yellow color. As CMY just looks a dirty grey due to ink limitations, equal amounts of CMY are replaced by pixels of black ink, which absorbs everything.

Pixels of two colors add together into a color I also see as smooth and continuous. These are the rules for subtracting the negative colors from white to get the full gamut of colors we expect from glossy magazines and book covers.

–C–Y = green   –C–M = blue   –M–Y = red

It is found, in practice, that varying the intensity of the RGB and CMY pixels in 256 increments—more than 16 million combinations—is sufficient to give the impression of 'full color' on both computer screens and print jobs.

These colors will not all fit these relationships on the real number line but the color relationships do fit perfectly onto the complex plane. It can be seen in the mismatch of the magenta arrows that these three will not fit into the same three-fold symmetry, but they do fit nicely into a construct with a hexagonal, sixfold symmetry.

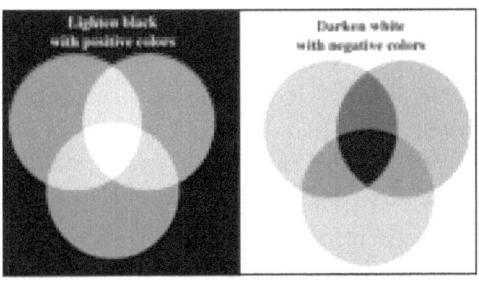

This can be simply illustrated using different modes in the image manipulator, Photoshop. In lighten mode, the overlap of three circles of positive color on a black background gives white and the three negative colors. In the darken mode, three overlapping circles of negative color gives black and the three positive colors.

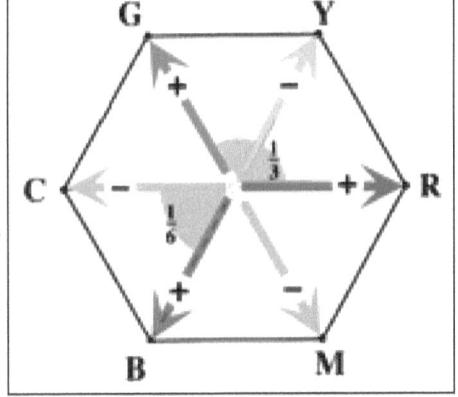

This a construct with a six-fold symmetry. It has each positive color and negative color along a plus-and-minus axis, and each axis has a rotation to the others of $\frac{1}{3}$ a period, with plus and minus ends alternating at a period of $\frac{1}{6}$.

Each positive color is embraced by the appropriate two negative colors, and each negative color is embraced by the appropriate two positive colors.

As an example of the very "unreasonableness" of one area of mathematics applying to two very different levels of the physical hierarchy, this simple math of positive and negative colors with its hexagonal, alternating symmetry of three axes rotated by $\frac{1}{3}$ a period and with $\frac{1}{6}$ a period separating them will be all that we need to understand quarks, gluons and the structure of the atomic nucleus (where the colors are called $\pm R$, $\pm B$ and $\pm G$, and CMY is not used). Colorless is important in this area, and we should note that adding red and anti-red to black gives colorless white, while subtracting them from white gives a colorless black.

It is for this reason that I have gone into the subject of color resolution in some detail.

When the pixels of color are small it is not possible to resolve the colors and the result is a colorless situation of white or black, or if all six colors are present in equal amounts, a colorless 50% grey.

Resonance

Resonance is phenomena that occurs with waves but not particles, and we shall encounter it a lot in the discussion. Waves that resonate together have a lower energy than when in isolation.

A relatively familiar example of paramount importance is the resonance of 4 different electron waves in the carbon atom resulting in the four 'hybrid' waves that account for carbon's chemical properties.

The four hybrid orbitals have less free energy than that of the single s-wave and three p-waves.

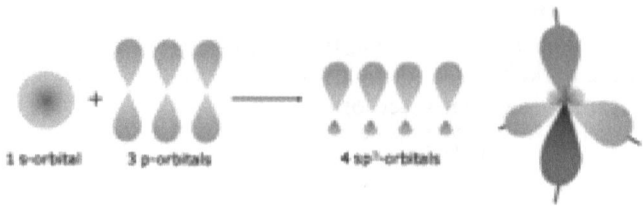

A Pixelated World

We will mention just three external aspects that only appear to be continuous because of resolution limitations:

Space

Space feels continuous, we certainly don't feel any 'grittiness' even when flying through the air. It is not, however, and it only appears continuous because our senses cannot resolve the grit. This 'pixel of space' is called the Planck Length, pL, and 'natural' units this size of pL is 1, while in human-scale units a pL is about a trillion, trillion, trillionth of a meter, 10–35m. Our best experimental instruments have a resolution of only about 10–18m, so space seems continuous even in the most delicate experiments—the pL emerges from theory and not from experiment.

Time

The sense of time passing continuously is even more ingrained, the concept of time being actually jerky is almost incomprehensible. As we pass through space (very slowly) and time (very rapidly), we cannot resolve the ultra-tiny 'pixels of time', called the Planck Time, pT, which is 1 in natural units and 10–44 seconds in human units.

Speed of Wave

We have discussed waves in the p-metric, and waves have a linear velocity as well as an angular phase. We shall now focus on the linear velocity of the wave and ignore its waviness. The velocity of waves in the p-metric is unchanging, it is a constant unity that never varies.

It was Einstein who first saw how velocity through time and space were connected in a 4-D world of spacetime that was an expression of the p-metric. The basic equation of Einstein's 'special relativity'

$$c^2 = (t)^2 + (xi)^2 + (yi)^2 + (zi)^2$$
$$= t^2 - x^2 - y^2 - z^2$$

is a simple extension of the Pythagorean Theorem. It states that sum of the squares of the velocities along the three imaginary axes (x, y and z) and one real axis (t) of the p-metric always sums to the square of lightspeed, which is unity, in natural units where $c=1=1pL/1pT$.

As noted earlier, while $+1$ and -1 have distinctly different properties, there is only the 'point of view' difference between clockwise and anticlockwise to distinguish $+i$ and $-i$. Moving in opposite directions along the complex time axis is the very significant difference between matter and antimatter. Moving in opposite directions along the imaginary spatial axes is the 'point of view' difference between going East and going West. This is why time is assigned the real axis and space the imaginary axes in special relativity nowadays rather than the space-is-real and time-is-imaginary of the early formulations.

To simplify the discussion it is no matter to orient the spatial dimensions so that the velocity is along a single axis, which we will call 'v'.

In the physical universe, the constant velocity of the wave can be distributed differently between space and time. Almost all the entities found in nature fall into just two classes:

All space and no time. An entity with no inherent rest mass, such as a photon, the velocity through space is 1 (lightspeed) and the velocity through time is 0.

$$c^2 = t^2 - v^2$$

Zero restmass

$$0^2 - v^2 = c^2$$
$$vi = c$$

All time and no space. For entities with a rest mass, such as things composed of real electrons and quarks, the real energy (mass) has an inertial preference for moving through real time and reluctance to move through imaginary space (as will be discussed later). A velocity of a 1,000 mph is considered 'fast' in our reference frame, our resolution of space and time. In natural units, however, the velocity is one millionth and the square is a trillionth, essentially zero.

Non−zero restmass

$$t^2 - 0^2 = c^2$$
$$t = c$$

As humans belong to the second class, we customarily move at well below lightspeed. This is where another artifact of resolution influences our perception of the universe. The phrase 'quick as a wink' suggests that 1/10th of a second in considered a very short time, while normal events take place in brief seconds and 'just a few minutes' and long leisurely hours.

Things look very different when viewed through special relativity. In a wink of an eye, the movement through time is equivalent to 18,000 miles in space. The passage of a leisurely hour is equivalent to a voyage of 670,616,629 miles, of a trip past Mars to Jupiter.

Our experience of smooth time passing slowly by the second is an artifact of resolution as it is actually jerky and exceedingly fast.

EXISTENCE

While the notion of 'existence' is usually considered a philosophical question, modern science has a precise description of it (just as the words 'internal and external' have the precise descriptions, 'measured by continuous complex numbers' and 'measured by integer real numbers' respectively. Existence, as defined by science, is also pixelated, it comes in discrete 'quanta' called Planck's Constant, pC or h, that, unlike time and space, is within the resolution of experiment. The pC involves a measure of 'energy-in-time called 'the action.

The Action

Max Planck is considered to have started the second scientific revolution when he discovered that 'the action' came in discrete quanta that were all the same size (the first, the start of classical science, is ascribed to Isaac Newton).

The seemingly smooth continuity of existence is another example of the resolution as the pC is so very tiny. The pC has a value of 1 in natural units while in human units it has the value

4.136×10^{-15} eV secs. This applies to linear considerations of existence, for angular considerations the pixel of existence is $h/2\pi$ and 0.658×10^{-15} eV secs, respectively.

A 100 kilo ($\sim 4 \times 10^{37}$ eV) man sitting and pondering existence for 10 seconds has $\sim 4 \times 10^{38}$ pC of existence, of pixels of existence, so it is no wonder he ends up thinking that reality is continuous as the jerkiness of existence is not resolvable, at least when unaided by machines.

The action is a measure of the Physical realm that plays a central role in modern physics. "Our search for physical understanding boils down to determining one formula. When physicists dream of writing down the entire theory of the physical universe on a cocktail napkin, they mean to write down the action of the universe. [The accompanying illustration is a contemporary action equation; 's' is the total action.] It would take a lot more room to write down all the equations of motion… The action, in short, embodies the structure of physical reality."[7] This is as true for quantum physics as it is for classical physics.

$$S = \int dx \sqrt{g} \left[\frac{1}{g} R + \frac{1}{g^2} F^2 + \bar{\psi} \not{D} \psi + (D\varphi)^2 + V(\varphi) + \bar{\psi} \varphi \psi \right]$$

Action and probability amplitude

It is an action equation that describes the combined influence of all the many interactions along an external 'history' of the entity to the changes to the internal wave at the end of that history. It is the action that connects changes in internal realm of the probability amplitude with the external interactions of an entity over a 'history' in time.

We have already encountered a touch of calculus when we discussed the area under a wave, the integral of the wavefunction intensity. The math connection between the internal and external is almost as simple. The line-integral of a wave is simply the linear-length of the wave if it was stretched out flat. For simple forms, such as a single cycle of radius r (a circle), the area-integral is $\pi r2$ and the line-integral is $2\pi r$. For intricate forms, however, while the area-integral is a relatively simple to calculate, the line-integral usually takes a lot of sophisticated math; we will not pause to explore it in any detail.

It is the line-integral that gives the total action along a history, and it this action that determines the final probability amplitude and intensity for any history. "The fundamental law of quantum physics states that the probability amplitude of a given path being followed is determined by the action corresponding to that path."[8]

It can be just by inspection, that the line-integral length of a sine curve goes up as the frequency goes up. There is more action along a high frequency wave than there is along a low frequency wave, a point we will return to shortly.

The Uncertainty Principle

To be 'real' then, as least as far as external science is concerned, an entity has to have at least 1 pixel

---

7 A. Zee, Fearful Symmetry, The Search for Beauty in Modern Physics, Macmillan, NY (1986), pp. 106 - 111.

8 A. Zee, Fearful Symmetry, Macmillan, NY (1986), p. 142.

of existence, there has to be at least one 1 h of energy-seconds. Existence comes in integer units of 1h, 2h, 3h, 1038h; never as ½h, 1⅓h , etc.

This is called "Heisenberg's Uncertainty Principle" when it is expressed in terms of observation and measurement: The uncertainty, Δ, of energy-in-time is the discrete Planck's Constant, the pixel of existence: $\Delta E \cdot \Delta t = h$. Another relation is that of momentum, mv, and position, x, where $\Delta mv \cdot \Delta x = h$. Attempting to resolve the physical world in greater detail, at say h/10 is futile, you can only get integer values of the pixels.

This is the insubstantial foundation upon which the Physical realm is constructed, as things are currently understood. The innocence and simplicity of matter-only classical science is gone, never to return, to the disappointment of those attached to classical science.

"Perhaps, someday, an experiment will be performed that contradicts quantum mechanics, launching physics into a new era, but it is highly unlikely that such an event would restore our classical version of reality. Remember that nobody, not even Einstein, could come up with a version of reality less strange than quantum mechanics, yet one that still explained all the existing data. If quantum mechanics is ever superseded, then it seems likely we would discover the world to be even stranger."9

Real and Virtual

Fundamental entities that have at least 1-pixel of existence are called 'real' particles. (The overloading of the word 'real' in math and science is not of my doing.) This requirement has different implications when applied to the two different basic types of fundamental entities: the bosons and the fermions.

Bosons

Bosons are easy to create. A +1-spin boson can be flipped away by a fermion which, by Newton's third law of Action and Reaction which holds in all sciences, flips by −1 into the negative state. A +½ rotating fermion that spits out +1 rotating boson will flip into the −½-rotation state—and the asymmetric wave is always doing this. When this flips out another +1 boson, it returns back to the +½ state (rotation add around the unit circle).

If this rotation is in both the complex wave of the p-metric as well as in the real topological defect, then the particle is a 'real' particle. In such a case, the fermion spin loses energy and the boson slips away with a pixel of existence. Fermions can rid themselves of surplus energy in the form of real bosons.

If the rotation of the fermion is only in the p-metric and not in the external space-time, the internal rotation of a fermion sends out 'virtual' bos-

9 Johnjoe McFadden, Quantum Evolution, W. W. Norton, NY (2000), p. 219

ons that only exist in the internal p-metric but not in the external space-time. They do not have a pixel of existence, and the fermion does not lose any energy.

Technically, a virtual boson does not have energy (which is always positive) but it does have momentum (which can be positive or negative). This is why a fermion can shed an unlimited quantity of virtual bosons—at a rate limited only by how fast it can rotate—sending them off in all directions in equal and opposite amounts to conserve momentum.

	FERMION ROTATING	BOSON EMITTED	
Internal	$+\frac{1}{2} \Rightarrow -\frac{1}{2}$	$\Rightarrow 1$	***Real***
External	$+\frac{1}{2} \Rightarrow -\frac{1}{2}$	$\Rightarrow 1$	***boson***
Internal	$+\frac{1}{2} \rightleftarrows -\frac{1}{2}$	$\Rightarrow 1, 1, 1...$	***Virtual***
External			***bosons***

A fermion is never in isolation, it is surrounded by a 'halo' of virtual particles at all times and places. It is actually a composite entity. This halo is the cause of what science calls the 'charge' on a fundamental entity, examples being the electric charge of an electron or the 'color' of a quark.

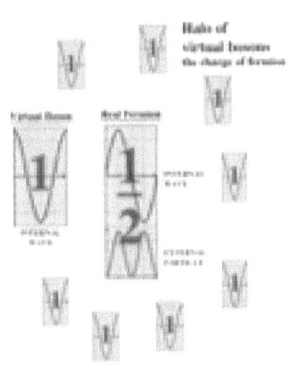

Virtual bosons, of course, being virtual and without any actual energy, are very hard to detect directly. They can, however, be indirectly detected because they can interfere either constructively or destructively, and the consequences of this are observed in the 'lines of electric and magnetic force' of classical science.

That the vacuum is awash with such "not really existing" virtual photons, is illustrated by the well-documented Casimir Effect of two silvered plates being pressed together by the 'vacuum fluctuations' when they get very close to each other.

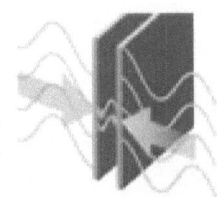

This happens because the virtual photons in the vacuum can have any wavenumber while those between the two conducting plates must have wavenumbers that fit the space. So there are a lot more virtual photons on either side than there are between the two plates. It is this imbalance that pushes the plates together to create the Effect.

Fermions

Fermions are not as easy to create as the bosons, because they are limited by their non-oriented twists. By the Law of Action and Reaction, these can only be generated from integers in pairs whose twists are in complementary directions along all three possibilities.

If a fermion-pair involves both the internal and external aspect, the fermions have (at least) a pixel of existence and they both have a real energy. For example, a high energy 'gamma' photon (a 2-twist boson) is quite stable when traveling through the vacuum of space but if it hits a "disturbance in the Force"—such as the electric field about an atomic nucleus—the pho-

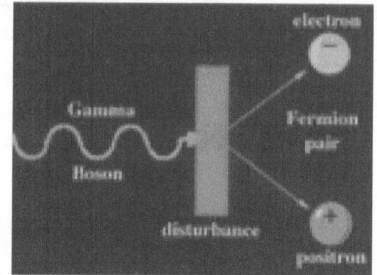

ton can fall apart. The boson turns into a pair of oppositely rotated 2-twist fermions; it creates an electron and an anti-electron (positron) that go their separate ways. This 'pair-production' of matter-antimatter out of gamma photons is the reverse of what happens when electron and positron merge their complementary non-oriented twists into integer-twist bosons, and gamma photons emerge in the mutual 'annihilation' of matter and antimatter.

In exactly the same fashion, a disturbed ultra-high energy photon can fall apart into three quarks and three anti-quarks with equal and opposite color and electric charges, a 'nucleon-pair' such as a proton and an anti-proton. The reverse happens in annihilation.

Antimatter is extremely uncommon in the Physical realm; any positron that has the temerity to appear in the Universe is rapidly annihilated by an electron to photons. (We will discuss this asymmetry of matter and anti-matter when we finish discussing fundamental entities and their interactions, and return to the discussion of the Hot Big Bang.) This is why matter is so long-lasting: there are no exactly-equal complementary twists to annihilate the electrons, which are 2-twist fermions, as the exactly-equal positive charge in the Universe is tied up in quarks, and they are 3-twist fermions. The quarks are eternal for the same reason.

This pair restriction means that virtual fermions always occur in matching pairs. An electron has virtual electron-pairs and bosons in its halo, but the pair effects are minimal because they do not have a pixel of existence. As the energy of a particle-pair is real, they can only appear briefly before annihilating back to nothing. For instance, the electron has a 'rest mass' of 511,000 eV so an electron/ positron pair has a real energy of 1,022,000 eV. Such a virtual pair existing for 4x10–21 of a second would have a Planck's Constant of existence, which they don't, so they have to get back together before this time. Even at the speed of light, one cannot get very far on a there-and-back journey in such a brief moment. The halo of virtual electron-pairs about a real electron only extends for about 6x10-13 meters, and only the highest-resolution experiments are capable of detecting its influence.

The Vacuum

As noted earlier, zero is an integer, and the undisturbed p-metric, the 'vacuum', has a probability amplitude, albeit a small one, to fall apart into a a fermion-pair that separate, albeit briefly, before recombining back to the undisturbed state. The vacuum is speckled with brief sparks of electro-positron pairs. The vacuum also has a prob-ability amplitude, albeit even smaller, to fall apart albeit even more briefly, into proton/ antiproton, neutron/antineutron pairs as well as all the other particles discovered by high-energy physics.

$$0 \Rightarrow +\frac{1}{2} \, \& -\frac{1}{2} \Rightarrow 0$$

A problem arises because fermion-pairs appearing in the vacuum have a rest mass of a million eV while they flicker on and off. The probability of a pixel of spacetime turning into a particle-pair is very small, and they only flicker very, very briefly (for arguments sake we will take the probability of a pixel of vacuum turning into a million eV electron/positron pair in any second the very small one-over-a-google, 10–100. The problem arises because even a very, very, very small number such as this can amount to a large number when multiplied by a truly gar-gantuan number such as a google-squared. The only exception is exactly zero. This conquers any number, no matter how large, such as a google-to-the-power of a google to the power google.

$$\frac{1}{10^{100}} \times 10^{197} = 10^{97}$$

$$10^{97} \times 10^{6} = 10^{103} \, eV \, / \, ly$$

$$0 \times 10^{100^{10^{100^{10^{100}}}}} = 0$$

If you do the math, the number of pixels in a cubic-lightyear-second is almost a google-squared at 10197, so the energy each second in a cubic lightyear can be expected to be over a google eV. This is just for electron-pairs; all the other pairs contribute as well.

Furthermore, there are a lot of cubic lightyears of vacuum out there— our Local Group of galaxies alone occupies about 1020 of them.

The gravitational effect of all this 'vacuum energy' in the virtual haze should compact space by a 'cosmological constant' factor of ~10100. This expectation of quantum physics is not borne out by experiment which has that the observed cosmological constant is essentially unity, there is essen-tially a zero vacuum energy in the Universe.

This calculation is based on a consideration of the p-metric alone, how-ever, and when the energy of the virtual particles in the s-metric are taken into account, the contributions of the two metrics cancel out almost com-pletely. As we shall see in a later chapter, the energy of real particles in the s-metric is negative, the dark energy that was only recently discovered to be a major component of the Universe.

Positive energy has a positive gravitation; it pulls together. Negative energy has a negative gravitation; it pushes apart. As plus a tiny number minus the same tiny number is zero, and zero conquers any number, no matter how large it is. The contributions of the equally-probable virtual 'supersymmetric particles,' as they are called, cancel out the contribution of the regular virtual particles, giving a dynamic cosmological constant at the observed size of unity and the observed zero vacuum energy.

$$\left(+\frac{1}{10^{100}} - \frac{1}{10^{100}} \right) \times 10^{197} = 0$$

$$0 \times 10^{6} = 0 \ eV/ly^{3}$$

Putting aside these global considerations for now, we will now return to the very local consideration of the real elementary entities out of which the Physical universe is constructed. We will take a closer look at each entity in turn, starting with the bosons, then the fermions, and then deal with their interactions with each other.

THE BOSONS

The bosons are the bits—more technically 'are the vectors'—of the fundamental forces that act on the fermions, the bits of matter. All things being equal, the symmetrical bosons of force are relatively ephemeral, and they easily appear and disappear. The asymmetrical fermions, in comparison, are relatively eternal, and only with difficulty appear, and then only in exactly opposite pairs of particles. The fundamental forces are:

The Weak force. The vector bosons are the W particles.

The Electromagnetic force. The vector bosons are the photons.

The Strong force. The vector bosons are the gluons.

Gravity. The graviton is thought by some to be the vector boson (an integer spin of 2) but, as we shall discuss in the next section, gravity is a consequence of wavefunctions being global, not local, on the p-metric.

We shall discuss the vector bosons in turn.

W-bosons

There are three W-bosons, the W+, the W–, and the W. This last is usually called the Z-boson, but as this obscures the basic similarity between the three bosons, we will not use it.

The W-boson has 1 unit of quantum spin, an oriented open wave with an enormous boundary energy of 91,200,000,000 eV, or 91.2 GeV.

Name	1	2	3
W ↕	1	0	0
W^-	$+\frac{1}{2}$	$-\frac{1}{2}$	0
W^+	$-\frac{1}{2}$	$+\frac{1}{2}$	0

This energy is reduced somewhat by transferring a ±½ twist to the second orthogonal direction. The result of this transfer is similar to an electron (or positron), as we shall see when we get to the fermions, but the magnetic spin is only ½ and it is going in the opposite direction. The boson resonates between the W and W± form, and this reduces the boundary energy of the W± boson to 80.4 GeV.

The real W-bosons are readily generated in high-energy particle accelerators, but their energy is so great that they quickly decay in fermion/anti-fermion pairs.

The virtual W-bosons are a component of the halos of all the fermions, but their enormous energy only allows them to exist very briefly so the virtual halo does not extend very far. A virtual W of 91 GeV for 10–25 secs would have a pixel of existence, so its lifetime is less than that, and it cannot get far even at lightspeed, so the virtual halo reaches less than 10–17 meters. A virtual W+/W− pair of 160 GeV has an even smaller halo.

These sizes are tiny, even on the scale of the atomic nucleus, and the fermions have to be extremely close, which is very unlikely, before their halos overlap and allow for interaction. It takes a long time, on a nuclear scale, for the weak force to play any effect. This is how the Weak nuclear force got its name. A neutron can decay, for example, into to a less-energy state of a proton, an electron and an anti-neutrino but this involves a W-boson intermediate so the   'half-life' of the neutron is ~11 minutes, which is akin to eons in nuclear-scale time.

The low probability of 'weak coupling' does play a very important role in everyday life as it this slowness that is responsible for the stately rate of thermonuclear reactions in the Sun over billions of years.

The half-life is a widely-used measure of probability when the probability of the change is constant over time. The half-life is the time it takes for 50% of a them to make the change. This is akin to measuring the probability of a regular coin by flipping it enough times so that, by the LLN, the desired accuracy is obtained.

Even though it is impossible to tell when an individual neutron will decay (there is a small probability of it lasting a second or a whole year, as there is of throwing 10 heads in a row), if unlimited numbers are available it is possible to measure the half-life that is precise as required. As large numbers are not a problem with neutrons, this measure of probability is known from experimental observation to be 611.0±1.0 seconds.

The Photon

The photon particle involves two oriented twists at right angles to each other. The first twist is called "magnetic" and the second "electric." These are two open waves that might be expected to have an enormous energy in the range of the single-twist W-boson.

Name	1	2	3
γ	1	1	0

Exactly the opposite is the case because of a resonance between the two waves. The first wave is constantly transforming into the second, and the second twist is transforming into the first. When the energy of one is high, the energy in the other is rapidly changing. The rate of change is called the derivative in calculus The derivative of a wave is the steepness, or slope of a wave is given by moving a dot along the x-axis by an infinitesimal length, dx, and seeing how the dot on the y-axis changes length, dy. The ratio of these lengths, dy/dx, is the derivative of the wave.

Closed sines from Open cosines

The energy of the open magnetic cosine drives the creation of the electric wave as its derivative, the change in the electric wave is maximal when the cosine is maximal. Similarly, the energy of the open electric cosine drives the creation of the magnetic wave as its derivative, the change in the electric wave is maximal when the cosine is maximal.

It can be seen from the illustration that a cosine wave is flat at its maximum, the derivative is 0 when the cosine is ±1. The cosine wave is rapidly changing as it passes through zero, its derivative is at a maximum of −1 (a downward slope when the cosine wave is passing through 0 from plus to minus) and a maximum of +1 when the cosine is going from negative to positive. A more detailed analysis shows that the derivative of the cosine wave is a negative sine wave that is −90° out of phase with the cosine.

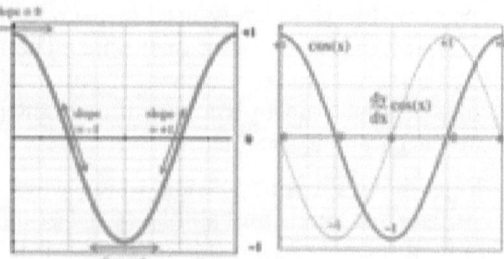

So the energy of the open magnetic wave is driving a closed negative sine wave in the electric, and the electric open wave is driving a closed sine wave in the magnetic.

The resonant form of two such resonating open

waves is two closed sine waves that are 90° out of phase with each other. There is no boundary energy in a photon, it has no 'rest mass.' A virtual photon is an electromagnetic wave with no inherent energy.

A real photon, however, has a pixel of action which is connected to the line-integral, and this goes up as the frequency goes up.

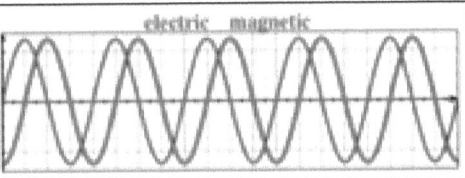

A photon whose wavelength is a light-year, has a very lazy wave with an ultra-low frequency, and has essentially no energy, the action is small energy and long time. A photon whose frequency is very high, on the other hand, will have a great deal of energy, the action is great energy and short time.

The magnetic and the electric are at right angles to both each other and the direction of travel. A photon that takes a long time, t, to cycle has little energy, E; a photon that takes a short time to cycle has a lot of energy. But the product of the two is always one pixel of existence, so for any photon, $Et = h$.

Being bosons in their behavior, photons readily link to each other creating what is called a ray of 'electromagnetic radiation.' Visible light is just 1-octave—when the frequency doubles— out of the 60 octaves of the entire spectrum. No matter what, though, each photon has exactly one pixel of existence, a Planck's Constant's worth.

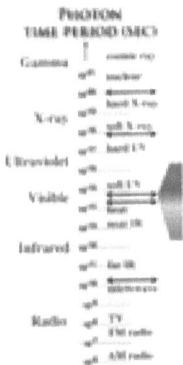

With increasing time-period, below visible photons there are the infrared, the microwave and the radio photons which fade out until their action is all in time, and their energy content is so small as to be undetectable. With decreasing time-period, above visible photons are the ultraviolet photon, the X-ray photons and the gamma-ray photons. The limit in this direction would be a photon with a time period of the Planck Time, the Planck photon we might call it with all its action in energy and the minimum of it in time. It is thought that such photons were present in the ultra-high temperature of the first few moments of Creation.

While the difference between an ultra-low frequency photon being used in submarine communication and an ultra-high energy cosmic-ray photon shedding a shower of real particles as it tears into the earth's atmosphere is extreme, they are all identical in that they all have a single pixel of existence, a Planck's Constant of the action—it's just distributed differently.

The table gives a few specific examples.

We have direct experience of these 'pixel of existence' photons. The light of a candle, for example, emits a wide variety of photons that peak in number at yellow light.

A yellow laser emits photons of identical frequency which, being bosons, all add together in lock step into waves of enormous size and power.

PHOTON	TIME PERIOD	ENERGY	ACTION
Radio	4.1×10^{-7}	10^{-8}	4.1×10^{-15}
Infra red	1×10^{-12}	4.1×10^{-3}	4.1×10^{-15}
Yellow light	1.3×10^{-15}	3.2	4.1×10^{-15}
X-ray	1×10^{-17}	4.1×10^{2}	4.1×10^{-15}
Gamma ray	1.3×10^{-21}	3.2×10^{6}	4.1×10^{-15}
Planck	10^{-44}	4.1×10^{29}	4.1×10^{-15}

All these photons are "real" in that they each amount to a pixel of existence.

Gluons

The gluons have three orthogonal oriented twists. If any two of these open waves try to put their boundary energy into driving closed sine

Name	1	2	3
g	1	1	1

waves in each other, the third one is going to be left out. This outlet for excess energy is blocked.

This is where the section on the color math of positive and negative colors comes into play. While the gluons involve many complex dimensions, the math collapses down to the simple math on the color plane we studied. The gluon open-wave energy is used to twist the p-metric away from its fourfold symmetry of orthogonal axes into a hexagonal symmetry of three axes. The amount of rotation is $\frac{1}{6}\pi$. The diagram illustrates that, of the thee axes, one axis is the most similar to the original real axis and we can call it the 'quasireal axis. The other two axes have little in common with the original ones. The quasireal axis is pointing forward in time while the other two are pointing backwards.

There are three spatial axes so there are three possible semi-real axes, a 'red' quasireal along the x, a 'green' quasireal along the y, or a 'blue' quasireal along the z axis.

In the hexagonal frame, the three non-oriented twists of a gluon fall along the three axes so that one is along the positive quasireal axis which moves forward in time and is the 'color' of the gluon, and the others fall along two of the negative axes, which combine as the 'anti-color' of the gluon.

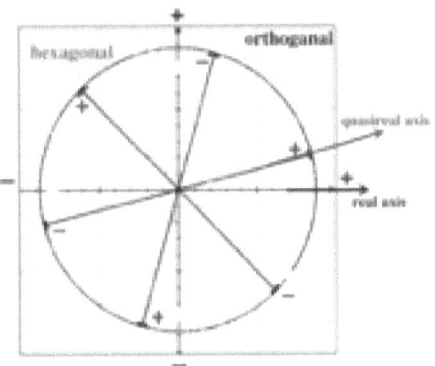

This is where the color-math of three positive and three negative colors simplifies things, where Y and M going in the opposite direction combine as –R, the C and Y combine as –G, and the C and M combine as –B. The result is a gluon with color and anti-color ends whose center is colorless.

The possible combinations are listed in the chart, where the gluons with an overall color are distinguished from those that are grey and colorless overall. A gluon can be considered as an entity with two colored ends with a colorless patch

	QUASIREAL		
	R	G	B
YM	+R–R	+G–R	+B–R
CY	+R–G	+G–G	+B–G
CM	+R–B	+G–B	+B–B

them. This allows the three non-oriented twists to coexist, but it does not solve the boundary problem of open cosine waves. A more detailed calculation shows that a gluon removed from a proton would have infinite energy, that the attempt at removal would take an exponential increase in energy without any bound whatsoever. Isolated 'color' is quite impossible.

The gluons get around the open energy problem using the limits of resolution. If the ends of a ball of gluons are such that the colors cannot be resolved in a pixel, it has no overall color.

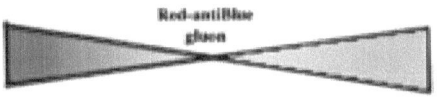

Red-antiBlue gluon

A ball of thousands of gluons aligned at their colorless centers into a sphere with the positive and negatives so mixed on the Planck scale that the surface is colorless has sufficiently reduced its energy so that can exist. The energy of all the open colored cosines is now in a colorless surface, a sphere of gluon-energy that has a colorless, energetic surface.

All the energy of the three open cosines is now in this colorless surface. A colorless state can be accomplished in two ways:

All the colors and anti-colors are equally represented, a grey of triple color pairs.

There is only one color and its anti-color in equal amounts, a grey of a single color pair.

The illustration shows these two types of "glueballs" as they are called, and a cross-section of the second

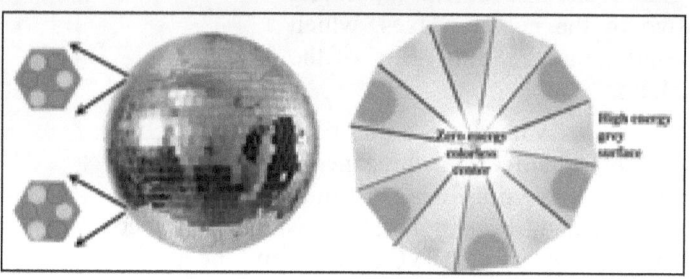

type, a glueball of +G–G (grey) gluons. The glueball has all its energy in the grey-of-triple color pairs surface and none at the colorless center. The surface of the glueball has the equivalent of a surface tension, akin to but immensely greater than, the surface tension that makes water ball-up on an oily surface.

Plain glueballs have a high energy of hundreds of millions of eV, and they are unstable and their energy readily decays into a host of real fermion-pairs and bosons. The glueballs are, however, stable when there are colored quarks in the composite.

Three quarks of different positive color (the nucleons, e.g. the proton) are surrounded by the a grey-of-triple-color surface. There is tiny imbalance of color over anticolor, but not enough to matter. A quark/anti-quark pair (the mesons, e.g. the pion) are surrounded by the single grey surface with no tiny imbalance of color over anti-color.

Symmetric in time

The final thing that we shall mention that holds for all the bosons is that they are open cosine waves, and these are symmetric waves, so they look exactly the same going in either direction along complex time. The bosons do not come in matter and antimatter forms (discussed in the next section). The complement of a +R–B 'matter' gluon would be the –R+B 'antimatter' gluon, but this is already present in the above accounting of combinations.

This concludes the brief discussion of bosons with their oriented twists. Next are the fermions with the non-oriented twists.

THE FERMIONS

All fermions are left-handed moving along the positive time axis in accordance with the inherent Left-handedness that we mentioned at the moment of Creation, when four complex planes were asymmetrically twisted apart into the p-metric and the s-metric The fermions fall into three classes determined by how many orthogonal ½-twists are in the p-metric.

The fermions having this left-handedness while moving in the positive (complex) time direction along the time axis are called 'matter' fermions and have a spin of $-\frac{1}{2}$. The fermions having this left-handedness but moving in negative (complex) time direction along the time axis are called 'anti-matter' fermions and have a spin of $+\frac{1}{2}$. While antimatter is moving backwards in complex time, it is the square of the complex value in the p-metric that gives the real-value in external spacetime, and both $+1$ and -1 when squared are positive. So both matter and anti-matter particles move together along the positive time axis in spacetime.

Three generations

As the chart illustrates, an electron can be thought of as a neutrino with second twist, and a quark as an electron with a third twist.

Name	1	2	3
Neutrino, v	$-\frac{1}{2}$		
Anti-neutrino, v̄	$+\frac{1}{2}$		
Electron, e	$-\frac{1}{2}$	$-\frac{1}{2}$	
Anti-electron, ē	$+\frac{1}{2}$	$+\frac{1}{2}$	
Quark, q	$-\frac{1}{2}$	$-\frac{1}{2}$	$-\frac{1}{2}$
Antiquark, q̄	$+\frac{1}{2}$	$+\frac{1}{2}$	$+\frac{1}{2}$

The neutrino, as discussed so far, has been considered a non-oriented twist that disorients a single spatial dimension, say x. Its full name is the 'electron neutrino,' for it is possible for the twist to disorient two spatial dimensions, say x and y, creating a muon-neutrino, or disorienting all three spatial axes, x, y and z, creating a tau-neutrino.

Adding the second twist to a muon-neutrino creates a muon, adding it to a tau-neutrino creates a tauon. Adding the third twist to a muon creates S-quarks and

	ELECTRON FAMILY		MUON FAMILY		TAUON FAMILY	
3		$-\frac{1}{2}$		$-\frac{1}{2}$		$-\frac{1}{2}$
2		$-\frac{1}{2}$		$-\frac{1}{2}$		$-\frac{1}{2}$
1	$-\frac{1}{2}$ 1-D		$-\frac{1}{2}$ 2-D		$-\frac{1}{2}$ 3-D	

C-quarks, adding it to the tauon creates B and T quarks. This creates three families of fermions that each come in three generations. Other than the

	ELECTRON FAMILY	MUON FAMILY	TAUON FAMILY
3	U/D quarks	C/S quarks	B/T quarks
2	electron	muon	tauon
1	electron neutrino	muon neutrino	tau neutrino

neutrinos whose rest masses are low, all the higher generations are high-energy and unstable, all decaying quite rapidly on the human scale. They were plentiful in the hot big bang, but play little role in everyday life.

Adding a second twist to the electron neutrino creates an electron, adding a third creates a D or a U quark. This is the First Generation of fermions.

Adding a second twist to the muon neutrino creates a muon, adding a third creates an S or a C quark. This is the Second Generation of fermions.

Adding a second twist to the tauon neutrino creates an tauon, adding a third creates a B- or a T-quark. This is the Third Generation of fermions.

The generations have all their properties in common except rest mass. The 0.5MeV electron, e.g. can be replaced by a muon (106MeV) or a tauon (1,778MeV) in an atom, and the quarks all have the same color properties and can replace those in a proton. The most massive fermion of all is the T-quark which weighs in at a whopping 173,200MeV.

As matter is composed solely out of 1st-generation fermions, we will only mention the other generations only occasionally.

Chirality

We have mentioned that matter is composed of fermions with complex left-spins of one-half (we will encounter a caveat to this later). The complex spin, however, is only the same as the observed real spin for entities that are moving at the speed of light. This places a limit on observation as you cannot overtake it, turn and observe it coming towards you.

We have already noted that clockwise and anticlockwise rotation are identical, they depend on if you are viewing the circular motion from front or back. As one can never overtake an entity moving at lightspeed, its complex spin and real spin are identical.

For an entity that customarily moves at sub-sub-lightspeed with a complex left-spin and real left-spin, it is easy to overtake it, turn and see the entity coming towards you with a right spin. A transparent clock goes in the anticlockwise direction if you

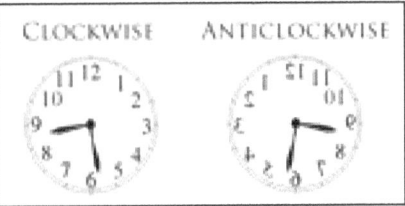

look at the back. (This is why +i and −i are so similar while +1 and −1 are not.) This is why entities with a rest mass inherent energy that move at sub-lightspeed through space seem to come spinning equally right or left. The fundamental left-handedness on the internal level was only uncovered recently by experiment, and its implications have yet to be fully understood.

The Neutrino

The simplest fermion is the (electron) neutrino, a single non-oriented topological defect that disorients a single dimension of

Name	1	2	3
Neutrino, v	$-\frac{1}{2}$		
Anti-neutrino, v̄	$+\frac{1}{2}$		

space. The left-spin neutrino moving backwards in complex time is the right-spin anti-neutrino. The size of the defect in space is on the Planck scale of 10–35 meters and, at the resolution of 10–18 in current instruments, a neutrino appears as a point.

This is a matter-antimatter pair of a left neutrino and right neutrino with negligible energy as they spin in the favored left sense of the p-metric. A right neutrino has never been observed, and calculations suggest that if it did, it would have an enormous energy, surpassing even that of the W-bosons.

The neutrino is a closed wave so does not have the boundary energy of the open-wave W-bosons. The misalignment energy of a few pixels of space is so minimal that the rest mass energy of the neutrino is only known to have less than 10eV and probably less than 0.01eV of rest mass

energy. The other generations are just a tad more energetic, the muon-neutrino at < 0.3 eV and the tau-neutrino at <31 eV. With such low rest mass, it does not take much to get neutrinos moving fast and at essentially lightspeed.

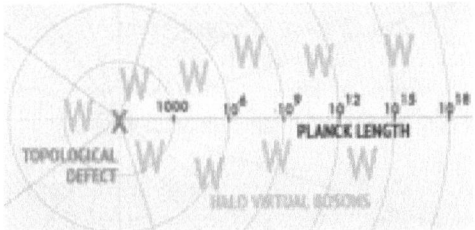

There is not much a neutrino can do, it is as close as anything gets in the physical realm to its abstract origins. Just about all it can do is to flip off virtual bosons with a single oriented defect, and this halo of virtual bosons is the 'weak charge' of the neutrino. This size of this halo about the pL-sized topological is 1017 pL, which sounds extensive but is only 10–18 meters, tiny even on a proton scale.

This is why the neutrino is said to be able to travel through a light-year of lead without being influenced at all, the chance that its weak halo will intersect that of

Name	1	2	charge
Neutrino	−½		Weak charge

another fermion in all that lead for any length of time is negligible.

The earth, naturally, is just as transparent to them and they flood upwards through us by at midnight as they do downwards through us at noon. This delight of nature inspired the writing of a poem about the neutrino by a major poet, and odes to fundamental particles are rare indeed. We are quite oblivious to this flux, but neutrino detectors are being constructed (underground ironically) as they provide an unparalleled view of the sun's incandescent core unobscured by the cooler layers it is massively swathed in.

The sun actually coverts ~2% of the energy generated in its million-degree core into a flood of of neutrinos, the remaining 98% is hard gamma photons. The neutrinos rarely have a close encounter as they leave the core at almost lightspeed and flash though the outer layers of the sun in two seconds and reach the earth in about 8 minutes.

While the sun is transparent to neutrinos, it is utterly opaque to all forms of photons. It takes an average gamma ray about a million years to make the journey

COSMIC GALL

John Updike

Neutrinos they are very small.
They have no charge and have no mass
And do not interact at all.
The earth is just a silly ball
To them, through which they simply pass,
Like dust-maids down a drafty hall
Or photons through a sheet of glass.
They snub the most exquisite gas,
Ignore the most substantial wall,
Cold-shoulder steel and sounding brass,
Insult the stallion in his stall,
And, scorning barriers of class,
Infiltrate you and me! Like tall
And painless guillotines, they fall
Down through our heads into the grass.
At night, they enter at Nepal
And pierce the lover and his lass
From underneath the bed – you call
It wonderful; I call it crass.

from the core, losing energy as it interacts along the way, until it is released from the 6,000° surface layer of the sun as a yellow photon which then takes 8 minutes more to reach the earth.

These detectors have even measured the tsunami of neutrinos released from an aging massive star when it becomes a supernova at a distance of thousands of light years. Betelgeuse, the pink star in Orion, is soon—in star terms—to go supernova and will become as bright as the full moon in visible light and flood the neutrino 'telescopes' with information about the first few minutes of the star's collapse.

The Electron

An electron is a neutrino with a second non-oriented left-twist added to it at right angles. A positron (anti-electron) is an anti-neutrino with a second non-oriented right-twist added to it at right angles.

While having a single non-oriented twist in the p-metric involves just a little 'rest energy' energy (~1eV), adding another non-oriented twist at right-angles places stresses on the p-metric.

Name	1	2	3
Neutrino, v	−½		
Anti-neutrino, v̵	+½		
electron	−½	−½	
positron	+½	+½	

These stresses can be relieved by resonating with another configuration. The misaligned energy of the second −½ spin of an electron can shift to boundary energy in a −1 twist of the neutrino axis. The electron reso-nates between 'two closed waves' state and the 'one open wave' state, and this resonance stabilizes the energy of the electron with a rest mass of ~500,000 eV. This resonance form of the

Name	1	2	charge
Electron	−½ ⇅ −1	−½ ⇅ 0	• Left spin • −1 electric charge ⇅ • Down magnetic spin

Name	1	2	charge
Positron	+½ ⇅ +1	+½ ⇅ 0	• Right spin • +1 electric charge ⇅ • Up magnetic spin

electron has a spin of −½, an electric charge of −½ (which is confusingly called −1), and a Down magnetic dipole.

The energy of an anti-electron is exactly the same, but the directions of spin are just reversed, a right spinning entity with a positive unit charge and an Up magnetic spin.

The electron is also surrounded by a halo of virtual bosons. The tiny halo of W bosons is the 'weak' charge of the electron and the extensive halo of virtual photons is the electromagnetic' of the electron.

Charge of electron

The two-twist virtual photons that are flipped off a two-twist electron have no rest mass; the wave travels in space but not in time. Unlike the time limit placed on the massive W-bosons, the virtual photons are eternal and have an unlimited range. While the halo of W-bosons is tightly confined, the halo of photons has an unlimited reach with their density falling off with distance.

In the ½ & ½ state, the electron flips off electric and magnetic waves that would wave in-phase with each other, unlike regular photons where they wave out of phase with each other. Instead, this 'spin polarization' sends the electric and magnetic components on separate courses.

The electric components combine as bosons to create the polarized 'lines of electric force' that radiate outwards from the topological defect. This aspect of the virtual halo is called the radial 'negative electric field' of the electron. The polarization of the virtual photons is the opposite of electrons.

The magnetic components combine as bosons to create the 'lines of magnetic force' that exit the topological defect from the clockwise side and return to the anticlockwise side. This aspect of the virtual halo is the N-S dipole 'magnetic field' of the electron.

Together, the electric and magnetic fields constitute the 'electromagnetic charge' on an electron.

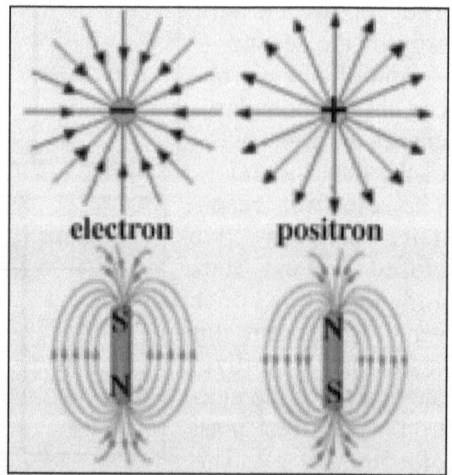

electron   positron

In the −1 state of the resonance, the electron flips off W-bosons, and this aspect of the virtual halo is the 'weak charge' on the electron. This is as ineffectual as the W-halo of a neutrino. Essentially all the interactions of an electron are when its electromagnetic halo intersects with the halo of another fermion.

The topological defect of the electron is on the Planck Length, its halo of W-bosons is almost as tiny, while the electric and magnetic are far reaching before the fields fade out towards zero.

This rest mass of an electron is 0.511 MeV, so it takes a gamma ray photon of over 1 MeV to create an electron/positron pair of real particles.

The Quarks

The quarks are entities with a third non-oriented defect at right-angles to both the second electron non-oriented defect and the first neutrino non-oriented defect. The state of three orthogonal ½-twists is of very high energy, and an isolated quark would theoretically have an infinite amount of energy.

Name	1	2	3
Neutrino, ν	−½		
Anti-neutrino, ν̵	+½		
electron	−½	−½	
positron	+½	+½	
quark	−½	−½	−½
Anti-quark	+½	+½	+½

As quarks are separated their energy goes up exponentially, and it was this separation of color charges by the inflation of the universe that caused the brake of the initial inflation into a hot big bang with an enormous energy density.

The state of three orthogonal ½-twists is of very high energy and this state decays into in lower energy state by shifting twist from the first onto the second one as the whole framework shifts into the hexagonal relation as discussed in the gluon section.

As discussed, this involves a rotation of the real axis by ⅙ into a quasireal axis in the hexagonal frame. This rotation can be accomplished in either of two ways with an electron and an anti-electron.

For the electron, a rotation of the −½ twist axis of the third defect by −2/6 in the favored left direction of the p-metric, shifts +2/6 onto the second −½ electron defect resulting in −⅙ defect on the electron level. This final state has an electromagnetic charge of −⅓ and is called the D-quark.

Name	1	2	3	charge
	−½	−½	−½	
D-quark	−½	−½+⅓ = −⅙	−½−⅓ −⅚	• weak charge • −⅓ electric charge • Color charge
	+½	+½	+½	
U-quark	+½	+½ −⅙ = +⅓	+½+⅙ = +⅔	• weak charge • +⅔ electric charge • Color charge

For an anti-electron, a rotation of the +½ twist axis of the third defect by +⅙ in the disfavored right direction of the p-metric, shifts −1/6 onto the second anti-electron $+\frac{1}{2}$ defect resulting in $+\frac{1}{3}$ defect on the electron level. This final state has an electromagnetic charge of $+\frac{2}{3}$ and is called

Name	1	2	3	charge
Anti D-quark	$+\frac{1}{2}$	$+\frac{1}{2}-\frac{1}{3}$ $= +\frac{1}{6}$	$+\frac{1}{2}+\frac{1}{3}$ $+\frac{5}{6}$	• weak charge • $+\frac{1}{3}$ electric charge • Color charge
Anti U-quark	$-\frac{1}{2}$	$-\frac{1}{2}+\frac{1}{6}$ $= -\frac{1}{3}$	$-\frac{1}{2}-\frac{1}{6}$ $= -\frac{2}{3}$	• weak charge • $+\frac{2}{3}$ electric charge • Color charge

the U-quark. The anti-matter U- and D-quarks are the same, except that all the signs are reversed.

This is where the earlier caveat about 'matter is made of matter fermions' comes as the U-quarks (which are by far the most numerous in the universe) actually have anti-matter components, but components so twisted that they no longer complement with electrons so they cannot annihilate, they can no longer untwist into bosons.

The 1st twist and the 2nd twist resonate as in the electron. The quark defect is surrounded by a tiny halo of virtual W-bosons, its weak charge, and an extensive but decreasing halo of virtual photons, its electric charge and magnetic spin similar to the electron and positron but in proportionate amounts.

Depending on the orientation of the quasireal axis in the x, y, z frame, the color charge of the quark can be +R, +G or +B, and the energy of this is got rid of by flinging off real gluons (and anti-gluons which are identical) and absorbing them. But gluons have a plus and minus color, and a color can only depart with a negative.

So a +R quark can fling off its color off by emitting a +R–B or +R–G gluon but, as a consequence, it becomes a +B or a –G quark. It can just as easily get rid of the +R by absorbing a –R+B or a –R+G boson and also becoming a +B or +G quark. The quark flings off its color onto gluons so quickly that its overall color falls below the Planck resolution and it becomes essentially colorless. When a +R quark emits or absorbs a +R–R gluon it remains unchanged, and it is incapable of emitting or absorbing the +B–G or –B+G gluons.

Three quarks, each a different color, can emit and absorb each other's colors quickly that all their color energy ends up in the blaze of real gluons, in a surface so pixellated with color and anticolor that it falls under the Planck resolution and appears as a hollow sphere with all its energy in a colorless surface. That lost amongst this haze is an extra pixel of red, green and blue is irrelevant to this glueball surface.

As fermions fit nicely in complementary-spinning pairs but not at all as triplets, so there are just two possible sets, the nucleons, of gluon-interacting three color-complementary quarks, the pair +U and–U with a D called a proton which has an overall electric charge of +1, or a pair +D and –D with a U called a neutron which has an overall electric charge of zero.

The total color charge of the proton, the sum of the UUD quarks, that has to be lost among the haze of gluon pixels is +⅔ +⅔ –⅚ = +½ while the color charge of the neutron with its DDU quarks is –⅚ –⅚ +⅔ = –1, and this difference shows up in the energy imparted to the glueball surface, the rest mass of the proton and neutron. The rest mass of the proton is 938 MeV while that of the neutron is ~0.1% at 940 MeV. This is sufficient to make the isolated proton a stable composite entity and for the isolated neutron to be unstable. As neutrons have to have more energy if protons are to be stable and result in an interesting universe, is a sign of the design of the Logos, the laws that govern the internal waves and the external result. If this were not so, the physical universe would be all neutrons.

The neutron decays into a proton when a D quark emits a W—boson and becomes a U quark. This real boson is so ephemeral that its mighty energy does not involve a pixel of action, and it falls apart into an electron and an anti-neutrino.

In composites with the other generations of quarks, the extra color is swamped by other considerations, and the 2nd & 3rd generation equivalents of the U-quark, the +⅔ C- and B-quarks, are more massive than the 2nd & 3rd generation equivalents of the D-quark, the -⅓ S- and T-quarks. The masses of the fermions are tabulated with the heavier quarks shaded.

	Family			
Generation	**1**	**2**	**3**	
1	Electron neutrino <2.2 eV	Electron 511 keV	D quark 4.8 MeV	U quark 2.4 MeV
2	Muon neutrino <170 keV	Muon 106 MeV	S quark 104 MeV	C quark 1.27 GeV
3	Tauon neutrino <15.5 MeV	Tauon 1.78 GeV	B quark 4.2 GeV	T quark 171 GeV

Just as it takes a disturbance of some sort for a gamma ray to decay into an electron/positron pair, it takes an overlapping of weak halos before the excess energy can be jettisoned as a real particle, albeit a very ephemeral one.

As far as size goes, the surface tension in the energy of the colorless glueball that surrounds the three quarks in a nucleon makes it a perfect sphere of size ~10–15 meters or 1030 pL. This is the size of nucleons.

A sphere that is 1030 pL across has a volume of 1090 cu. pL which gives the colorless quarks plenty of colorless space for them to roam in. The topological defects of a quark are on the Planck scale, and the accompanying halo of W-bosons only has a volume of 1050 cu. Pl. The colorless space available for the quark to move around is 1040 the size of the weak halo. This is akin to three peas in a cube with sides the earth-sun distance which is why the quarks hardly ever intersect. It is only when they oh-so-rarely intersect that the excess energy can be dumped into a W-boson and the neutron can beta-decay into a proton. The 11-minute half-life of the neutron, an eon in quark time, is the result of this 'weak decay.'

All the color energy of the quarks is in the glueball surface and the inside the hollow sphere is colorless. The quarks are also colorless, and they only interact with each other by their electromagnetic charges and magnetic spin. They have "asymptotic freedom" from the strong color force that is also called "ultraviolet slavery" for woe betide a quark that ventures to near the colored surface as its energy goes up exponentially.

If a quark were to be knocked out of a nucleon, the energy would suffice to make an antiquark/quark pair and the departing quark would be a quark-antiquark pair in a glueball, a ball of grey-of-grey gluons called a 'pion.' Pions can be uncharged, such as +U–U and +D–D with a rest mass of 135 MeV, or charged pion such as a +U–D with electric charge $+\frac{2}{3} +\frac{1}{3} =$ +1, or a +D–U with a charge of $-\frac{1}{3} -\frac{2}{3} = -1$. The charged pions have a 1% greater rest mass of 140 MeV.

The matter-antimatter combination is unstable and a negative pion decays in just 10–8 seconds into a muon and an anti-muon neutrino. The neutral pion has an even shorter half-life of 10–17 seconds and the quark and antiquark unwind into two photons that zip off in opposite directions.

At resolutions where a proton becomes a point, it behaves exactly like a positron that had put on mass from 0.5 MeV to 938 MeV. But electrically and magnetically, a positron and a proton are identical at an atomic resolution.

This brings us to the topic of the interactions between the fundamental particles. We have already touched on this as quarks cannot be discussed individually but only as colorless combinations. The quarks rid themselves of color by constantly creating a hollow glueball of complementary colored gluons within whose insurmountable walls they lead a colorless, if confined, existence of low energy. In the language we will establish in the next section, the quarks interact by 'coupling' with gluons, the aptly-named Strong fundamental force.

THE FUNDAMENTAL FORCES

Modern science recognizes only four fundamental forces in the Physical realm out of which all other forces are derived:

1. The Force of Gravity
2. The Weak Force
3. The Electromagnetic Force
4. The Strong Force

These four play very different roles in the functioning of the Universe. Gravity is a global entanglement that is responsible for the large scale structure of stars and galaxies. The weak force is an extremely local effect of W-bosons and is responsible for the stately rate of the thermonuclear reactions that power a star in its long middle age. The electromagnetic force involves photons and is far reaching, but as it tends to cancel out, is responsible for the local structure and chemistry of atoms and everyday matter. The strong force is very local, involving gluons, and is responsible for the structure of the atomic nucleus and the variety of stable and radioactive elements.

Gravitational interaction

Gravity is a result of the internal entanglement present in the p-metric at the beginning that binds everything together in the Physical realm. The energy that is the waves and distortions of external spacetime send out virtual bosons that, like photons, have an unbounded range. This is the internal aspect of what is called a graviton. It is a symmetrical boson, it has an oriented twist, but unlike the other bosons, it has a spin of 2. To return the same state when rotating in this takes only ½ a rotation. If you go the whole distance back to the start, you will have rotated twice.

Gravity is a global effect wherein the entangled bosons of the entire p-metric combine and blend into a single entity called the curvature of spacetime.Where they are abundant, spacetime is more curved; where they are scarce, it is less curved.

The entity emitting virtual gravitons, which like virtual photons carry momentum but no real energy, is also influenced by the gravitons that are flooding by.

Inertia and Mass

The influence on the internal wave of a rest mass energy by the internal aspect of gravitons is called 'inertial mass' and is the 'gravitational potential' of the spacetime being curved by virtual gravitons. The effect of this is to resist any change in momentum, the object has a linear and angular momentum that it takes energy to change. They are distortions on four separate complex planes, so their density and influence falls off with the distance, d.

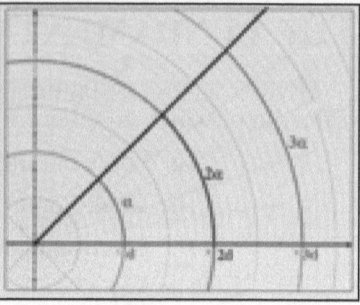

The influence on the external aspect of momentum is in external space and falls off as the square of the distance. This influence is called 'gravitational mass' and the overall influence is the 'gravitational field.' The external transfer of momentum between mass-energy entities causes them to draw together; there is a gravitational force between them.

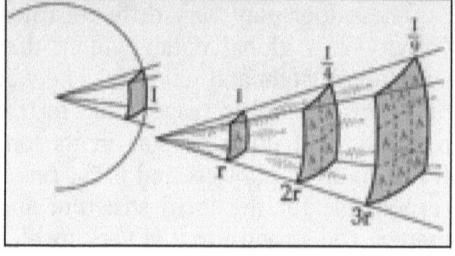

The virtual gravitons all blend together into the structure of spacetime, which is essentially flat except where it is distorted by large concentrations of mass-energy. And they do have to be in very large concentrations since the momentum carried by virtual gravitons is minuscule compared to that carried by virtual photons. The strength of the electromagnetic force is ten trillion-trillion-trillion times greater than the gravitational forces. The magnetic attraction of a weal refrigerator magnet, for instance, holds it firmly attached to the door, and the gravitational force of the entire earth pulling on it is insufficient to dislodge it.

The inertial influence of gravity falls off with distance while the mass aspect falls off as the square of the distance. This makes quite a difference.

To illustrate, consider a universe in which all the mass-energy is spread out at a constant density. The mass in a sphere of radius 1 and volume $4/3\pi3$ surrounding you has an inertial and a mass gravitational influence which we can both set at 1.

A surrounding sphere of radius 2 adds a greater volume, and the volume of the shell is the volume of the sphere minus the volume of the first shell. A second shell of radius 3 is the volume of the third sphere minus the volume of the second. Another shell of radius 4 has it volume minus that of the 3rd sphere. The inertial influence of each added shell falls off with distance while the mass influence falls off as the square of the distance.

In can be seen from the graph that the inertial influence increases rapidly as the shell gets bigger, while the mass influence falls to a constant. While the universe is most certainly not homogenous on a local scale, at cosmological scales the distribution of matter is essentially uniform, and resolution enters the picture.

The implication is that the distant galaxies play a greater role in the inertial mass than the closer ones do. This is why inertial mass involves the entanglement of all the mass in all the universe, as in Mach's principle of inertia. Momentum is a vector property where direction counts, and the global influence of the graviton entanglement imposes what is called the Conservation of Momentum. In any situation the overall momentum is a constant. The momentum before and after an interaction is the same. This is why a matter/antimatter pair with zero overall momentum annihilates into a pair of photons that speed off in opposite directions so that the overall momentum is still zero. This is another example of 0 turning into +1 and –1. A single real photon has momentum, so annihilation into a single photon would be unbalanced and create momentum, which is forbidden.

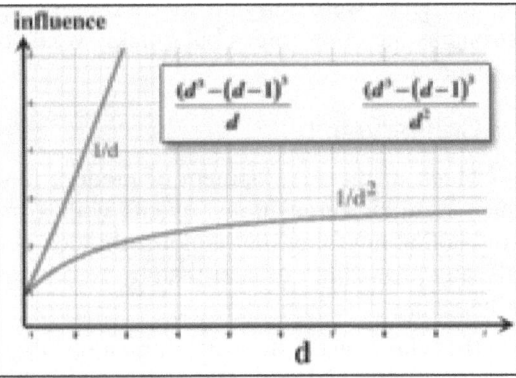

As noted, the velocity of the internal wave is a constant lightspeed. For entities with a rest mass, this velocity is almost entirely along the time axis and zero along the spatial axis. If energy of kinetic motion is added to an everyday object and its velocity approaches a substantial fraction of light speed, its velocity through time diminishes in proportion.

Acceleration through space involves rotating the direction of the constant velocity away from the time axis towards the space axis. For example, when this rotation is 45° the velocity through time and space are equal. The velocity through space is 0.71c — a speedy 473,478,700 mph—while the passage of time is 71% the rate of everyday objects, called 'time dilation.'

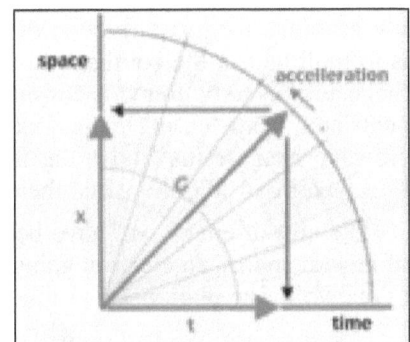

The energy added adds to the mass and inertia, such that as the velocity approaches the speed of light the mass increases without bound, and would be infinite at the speed of light. This is why objects with an inherent mass energy cannot travel at the speed of light. A proton moving at essentially the speed of light can have the inertial and gravitational effect of the planet earth, and greater.

In the maelstrom of energy in the first few moments of the hot big bang, all particles were moving at essentially lightspeed, and gravity was as strong as the other forces but quickly weakened as the Universe cooled and expanded.

Conservation

Much of the development of modern fundamental physics is driven by such forbidden transitions, events with a probability of zero. The golden rule is that anything that is not forbidden is compulsory; given enough trials, any probability greater than by zero by even a tiny amount will eventually occur.

It is the conservation of momentum that underlies Newton's 3rd   law that states that action and reaction are equal and opposite.

The gravitational influence is local, the influence of the distant galaxies is about the same as the nearby ones, and the pull from one side tends to cancel the pull from the other side. External energy, being the square of the internal momentum, is a scalar property where direction does not count, and the imposition of the entangled gravitons that energy can be neither created or destroyed, the Conservation of Energy, it can only change form and distribution in any interaction. This is called the 1st law of thermodynamics. The gravitational effect of energy is so very small that it takes a very large amount of it to be appreciable, which is reflected in the enormous planet-scale amounts it takes to be noticeable.

On a cosmological scale, gravity opposes the expansion of the universe, and the effect of all the mass in the universe is to slow its rate of expansion. We will return to this when we get to the topic of 'dark energy.' It is now known that all the visible matter in the universe, the stars and galaxies, accounts for just a fraction of the total energy in the universe; the rest is to be found in the confusingly-named 'dark matter.' In just what quanta this energy is to be found is currently unknown but is possibly in the 'relic-neutrinos' from the big bang. Like the 'relic-photons' of the 'cosmic microwave background' radiation, they outnumber the electrons and quarks by a factor of 100 billion, and their tiny masses all add up.

The unit of energy we have been using is the electron-volt, the amount of kinetic energy an electron gains as it moves across a potential difference of one volt. The photons of visible light have energy on this scale.

As there are two trillion trillion trillion eV of energy in a gram, and a gram of mass is considered small, an eV would be an awkward unit for everyday use. Energy in huge, concentrated amounts is called mass; the conversion between the two scales is Einstein's famous equation.

$$E = mc^2$$

The 2nd law of thermodynamics arises from the driving force of interactions called the Principle of Least Action, that entities interact so as to minimize the overall action created by energy-in-time. The inverse of this is called the entropy of a system and the 2nd law is often stated as the entropy of a system always increases to a maximum.

We observe the principle of least action as the tendency of all things to minimize their energy. To get rid of as much energy as possible. To move from an excited state to the ground state.

The principle of least action is at the very foundations of the Logos, of the laws of nature. This is why change in the physical world is said to be driven by energy transfer as all things seek to minimize their 'free' energy, the basis of thermodynamics.

Free energy simply refers to the energy that can be altered, given the circumstances. In chemical changes, for example, the energy that is tied up in the 'mass' of the entities is not able to change, only the energy in electron waves can alter. It is this energy, then, that governs the changes in the waves in chemical interactions.

The implications of the 2nd law are quite different for interactions involving gravity—which always pulls things together—and interactions that involve electromagnetism—which can pull together and push apart. The state of least action and maximum entropy for gravity is when interacting things are all concentrated together, while for electromagnetism the least action is when the interacting things are smoothly spread out.

It is a balance between these opposing tendencies that keeps a massive star in an extended state while it is generating photons during its long maturity, and the gravitational collapse that occurs when the ability to generate photons eventually ceases.

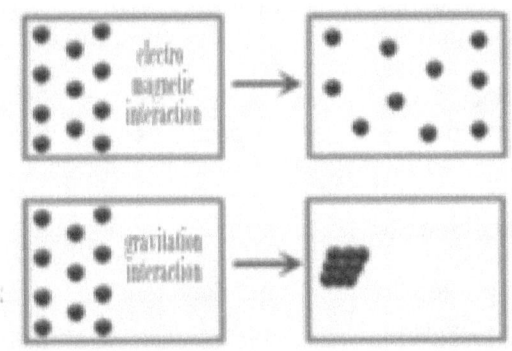

Quantum gravity

It must be noted that the current understanding is still an open question as no way has yet been found to link the local and global aspects of gravity into a unified view.

The pixels of a 'perfect vacuum' have an internal wave aspect that is usually cancelled out by all the others around. Its quiescence is only barely disturbed by the virtual particles it very occasionally turns into. The pixel also has an excited, but not twisted, state. This high-energy state is called the Higgs Boson, and it has no spin, no charge, and no color. A search is currently on to detect a Higgs, and the current lower bound for its rest mass energy is ~160 GeV, about 160 times the energy in the gluon field inside a proton. A virtual Higgs would have an influence even more limited than the W-bosons, and it is thought that it is coupling on the Planck scale with virtual Higgs that causes energetic defects in the p-metric to radiate gravitons.

So just like the other three forces, gravity on the ultra-local level can be described as coupling with virtual Higgs bosons, and as coupling with gravitons on a local level.

This local aspect of gravity is the quantum description of gravity.

On a global level, spacetime can be treated as a continuum, and the mathematical tools dealing with the curvature of a smooth continuum are well developed. This perspective was pioneered by Einstein who described the global aspect of gravity as a bending of spacetime in his epochal General Theory of Relativity.

Unfortunately, the local quantum description and that of global General Relativity have yet to be reconciled. Richard Feynman attempted such a unification when he considered the link between local coupling with spin-2 gravitons and the global coupling with spin-0 bosons. This was well before even the possibility of the Higgs had entered the scientific lexicon, and he did not pursue the concept. He did, however, perfect our current view of the electromagnetic force, the topic of the next section.

Electromagnetic interaction

When the virtual halo of polarized electric photons moving out around the electron (its negative electric field) intersects with the virtual halo of oppositely-polarized electric photons moving out around the positron (its positive electric field), they have the same sense of rotation. The spiral leaving the electron has the same sense as the spirals arriving from the positron, and they connect up. The magnetic photons of the electron and positron are in opposite directions and they tend to cancel each other out.

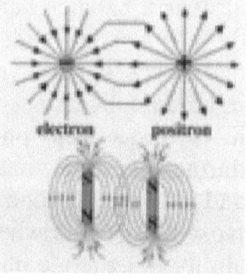

The external result of the momentum exchange via the virtual photons is to move the waves of the two particles together, just as gravitons do.

When two electrons intersect their fields, the leaving and arriving virtual photons have the opposite spins, and they cancel rather than augment each other. It is the magnetic photons that are going in the same direction and the magnetic 'lines of force' are bunched up together.

The external result of the momentum exchange via the virtual photons is now to move the waves of the two particles apart, which gravitons with their 2-spin non-polarizable state do not do.

This influence of virtual photons is the cause of the observation that like electric charges and magnet poles repel each other, and that opposite charges and poles attract each other.

electron   electron

This cushion of virtual photons can be directly experienced by attempting to unite the N poles of two strong magnets. The experience of the invisible cushion keeping them apart is about as tangible as virtual photons get.

Atoms

When an electron and positron are near, their probability amplitude waves combine into a single resonant standing wave that is called an 'atomic orbital.' The probability density of them within this standing wave is equal in a sphere of about 10–10 meters—small by human standards but vast on the scale of electrons and quarks.

This composite entity is called an 'atom of positronium,' and it is another example of 'system building,' when simple systems interact together to create composite systems. Three quarks interacting together to create a proton is another example.

In the simple terminology we will find useful, positronium is a composite system of two subsystems (the electron and positron) that are coupling with subsystems from their own structure (the virtual, connected photons). The system has an internal standing 'system wave' (the resonance of the internal probability amplitude waves of the electron and positron) and an external substantial form (the composite of the overlapping probability densities of the electron, positron and virtual photons). Even simpler, we will say that the internal system wave confines the history of the subsystems into an external form that reflects the internal form.

If the electron and positron happen to end up too close together, their complementary defects untwist into real photons and the positronium 'atom' decays into gamma photons. We earlier noted that a proton behaves exactly like a massive positron but with twists that do not complement those of an electron.

The combination of an electron and a proton is the stable hydrogen atom, and they both have a probability density in the same system wave. The proton, being 1800 times as massive as the electron, shifts only slightly and jitters around its probability density at the center of the standing wave very much less than a positron would. The light electron has the same probability density as in positronium.

In an atom in which the electron is replaced by a 3rd generation tauon, the tauon which is just as massive as the proton, quivers in much the limited space as the proton, and can be said to be 'orbiting' within the proton. The 'tau-atom' is neutral and it can get very close to other such atoms. The brief lifetime of the tauon is the only barrier to it being a catalyst for nuclear fusion.

The hydrogen system is composed of two subsystems that interact by coupling with their subsystems; it has an internal standing wave called the atomic orbital, and an external form that is a composite of the probability density. The atom is 'substantial' because of the electron cloud, it is massive because of the proton (which itself is massive because of the energy in the blaze of gluons).

The hydrogen atom has an internal system wave and an external system form of confined particles that reflects the form of the internal.

We have already seen that the system wave of a proton perfectly confines its quarks and gluons but imperfectly confines the virtual photons that spill out as the electric charge and magnetic dipole of the proton.

The system wave of a hydrogen atom does not perfectly confine the subsystems, and this imperfect confinement is responsible for the chemical behavior of the hydrogen atom. The orbital is capable of holding a pair of oppositely-spinning electrons in a resonant, low-energy state but it only contains a 'singlet' electron.

INTERNAL	EXTERNAL	
SYSTEM WAVE	SYSTEM FORM	
Resonance of:	Confined probability density of:	SYSTEM
1 electron wave	electron	Interacting subsystems
1 proton wave	proton	
virtual photon waves	Virtual photons	Coupling sub-systems

A example of an almost perfect confinement is to be found in the helium atom with a nucleus of positive charge two and a pair of opposite-spin electrons filling the orbital. The helium system almost confines its subsystems as well as a proton does its quarks. It has no imbalance to give it any chemical properties, and it is the closest that reality comes to the massy-spheres of solid matter considered primal in classical science.

Helium atoms behave almost exactly as classical science expected two tiny billiard balls to behave. This is looking only at the external aspect of interaction, but all the interesting stuff is actually happening on the internal level. The two helium atoms have nothing to gain by resonating together, and the waves bounce and each other and go off in the opposite direction. The well-confined quarks, gluons, electrons and connected virtual photons that are a helium atom preclude any interaction and they are so barely sticky that only a temperature near absolute zero will condense a gas of them into a liquid.

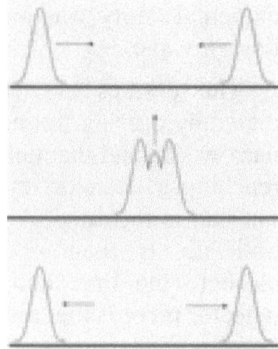

This is the source of the everyday experience that matter is solid, even though it is not. Two helium atoms colliding will elastically bounce off each other just like two tiny billiard balls, just as if they are solids. But it is actually the self-contained waves of virtual photons that fill each atom that are doing the bouncing, not anything solid. The mass aspect of 'matter' is the consequence of a blaze of gluons; the substantial aspect of matter is a blaze of virtual photons. They are emergent properties not fundamental properties. Such are the philosophical implications of modern science that are, as yet, only partially digested.

Quantum mechanics

The area of quantum mechanics that deals with the interaction of photons and electrons is known as Quantum Electro Dynamics (QED). Richard Feynman's book, QED, The Strange Theory of Light and Matter is an excellent overview of its triumphs. In QED, the internal aspect is called the 'probability amplitude' and is described and measured with complex numbers.

Quantum mechanics describes the way that waves of probability amplitudes combine and interfere by the arduous, if accurate, method of adding and multiplying thousands of complex numbers to calculate the final probability amplitude, p@a. The probability density of the particle aspect is then calculated by the elementary step of calculating p2.

Slit experiment

One of the earliest experiments that revealed that there was more to the physical universe than just the external aspect, and thus was instrumental in the emergence of quantum science that included the internal, was the slit experiment. This experiment was originally designed to determine if a fundamental entity was a particle or wave.

The entities are collimated by passing through a narrow slit, and the number reaching a detector on the far side is recorded. A second slit is then opened close to the first, and the detector records that result.

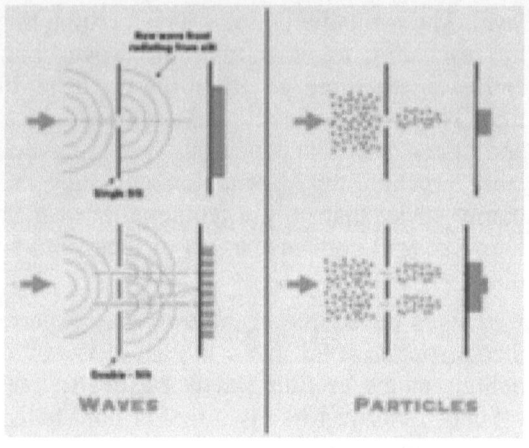

With waves, there is interference of the waves, and an 'interference pattern' is created at the detector. In particular, there are places that detected arriving entities with one slit open but detect nothing at all when two slits are open. This does not happen with particles, opening another slit can never stop particles reaching the detector through the first slit.

The slit experiment confirmed that light was a wave, as is easily confirmed by monochromatic light—all of one frequency, or color, such as the yellow generated by electrons jumping about in the orbital of sodium—when the 'up' amplitude of one wave can be exactly cancelled by the 'down' amplitude of the other wave.

We have seen that a simple wave is described by its amplitude and its phase, just where it is along the sine wave. This will depend on how far the wave has traveled, x, and its phase which will depend on is period of the wave, how far it has traveled, $\sin(t(x))$. When the length of one path to x differs from the length of the other path by ½ a wavelength, the waves will be exactly out of phase and interfere destructively. Conversely, the brightest places will occur at spots that differ in distance by exactly one whole wavelength.

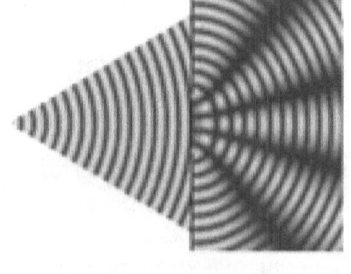

At first, it was thought that it was the external wave that that created the interference pattern, but then came the development of sources and detectors that could deal with

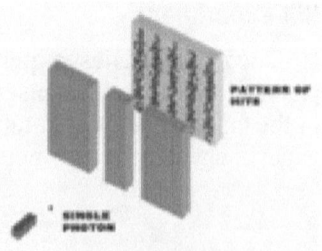

single photons at a time. The result was confusing. These single photons behaved like a particle in that they always arrived as a single spot and not spread all over the detector. But, as the pattern of spots built up one by one, an interference pattern emerged. The single particle-like photon was interfering with itself, and the wave was going through slits at the same time.

This was not an external wave, it was the internal wave that determined the probability of a photon reaching a spot on the detector. Such a wave did not exist in classical science, which caused endless confusion. The amplitude of the final wave at a spot on the detector determined the probability that the photon would interact there. This is how the phrase 'probability amplitude' entered the scientific vocabulary.

At first, the slit experiment seemed to show that electrons were particles. Then it was realized that the distance between the slits had to be commensurate with the wavelength of the wave, and that separations that were suitable for light would not work for electrons; only atomic size separations would work. The regular planes of atoms in a crystal, such as salt, are just this size, and electrons passing through these 'slits' also had a diffraction pattern, now a 2-D pattern (as illustrated). An electron also had a wave of probability amplitude associated with its particle. The electron also interfered with itself and seemed to pass through both slits at once.

The basic rules used in QED for calculating the final probability amplitude are simple and similar to those of classical probability.

If there are alternative ways of it happening, add the probability amplitudes. (Classical theory says add the probabilities.)

If a series of intermediate steps is involved, multiply the probability amplitudes for each step. (Classical theory says multiply the probabilities.)

The square of the magnitude to the final probability amplitude (the p in p@a) is a real number that is the probability, p2, of it occurring. (Classical theory omits this step—adding and multiplying real numbers always results in real numbers.)

All of quantum mechanics is just repetition of these two operations, just millions and billions of times. Much of the sophisticated math used in QED is just shortcuts to doing this endless iteration of adding and multiplying complex numbers.

This method works just as well for standing waves. If you add and multiply all the probability amplitudes for an electron and proton, the end result is the 1s orbital as calculated from the Schrodinger Equation. The adding and multiplying method is called the 'sum over history' method and it is mathematically equivalent to the differential wave formulation.

What we have described so far is sufficient to understand all nonliving systems. The form of internal wave gets more and more intricate, and the probability density of the interacting subsystems follows along. These intricate waveforms express more and more sophisticated emergent properties that are inherited from the Logos.

But the basic principle is the same: an internal wave determining the external form to the probability density of electrons and atomic nuclei.

Just how intricate a form can result from simply adding and multiplying complex numbers together? Luckily, research in this direction has already started, and the answer to this query would be, "Very." We shall take a brief look at these simple beginnings at exploring the properties of forms generated by adding and multiplying complex numbers.

Sophisticated internal forms

This possibility of exploring this aspect of mathematics only really opened when computers were developed which could add and multiply millions of complex numbers a second. (Some early pioneers of the field actually added and multiplied thousands of numbers by hand!)

As we saw earlier, the endless and fascinating forms of the Mandelbrot Set are likewise created by repeatedly adding and multiplying complex numbers, so we should not be surprised at the sophistication involved in everyday matter. If our computers were powerful enough, they could use the equations of QED to calculate the forms of ordinary objects.

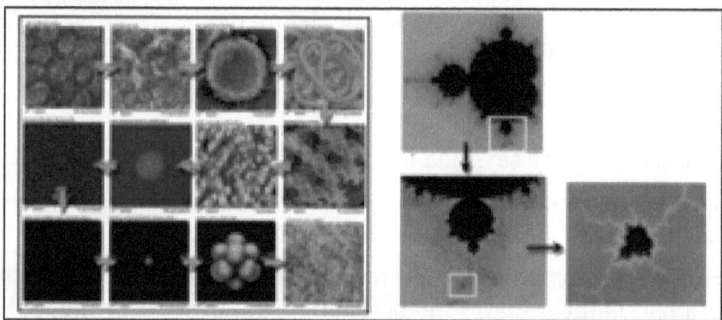

The illustration compares the forms that result when magnifying the Mandelbrot and a powers-of-ten zoom-in on a plant leaf showing how the patterning changes in a similar way. The photomicrograph is recording the result of photons from a source interacting with the electrons in the specimen, and this is just what QED is good at.

Julia sets

The above illustration of the Mandelbrot was generated by what is called 'serial computation,' each complex number in the chosen range is taken one at a time. The central processor then iterates it in the equation and, if the sequence that results is a dust or a bounded sequence, it instructs the graphic processor to color the pixel corresponding to the number appropriately. Then the next number is tested until all of the numbers in the range have all been serially processed in order and the results sent to the screen.

There is another way of generating the Mandelbrot set, and this is by 'massively parallel processing,' where all the numbers are tested at the same time. As mentioned, the numbers in the set create bounded forms on the complex plain, and the form that the number creates is the 'Julia set' for that number. The number 0 has the most boring of Julia sets, when you feed it into the equation, it just stays zero no matter how many times you iterate it.

Other numbers create bounded forms of intricate shapes, for example the six complex numbers in the illustration shapes.

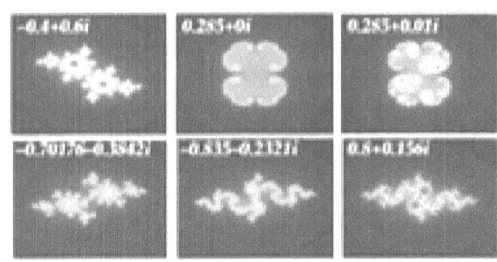

The parallel computer we will consider has a processing unit within each pixel—and this will have to be a thought experiment as current computers can manage only eight central processing units (CPU) at a time. Each computer is numbered by its position in the pixel array. Each pixel receives two pieces of information: the complex number it corresponds to, and the program to iterate the equation.

Each pixel now computes the Julia set for its complex number and writes its result to the display. While this parallel process is more sophisticated than the serial computer, the serial computer can be programmed to model the parallel computer. The result is illustrated for a small array.

The Julia sets are combining with each other into the Mandelbrot set. This is somewhat similar to the Fourier process already described where any shape wave can be broken up into into a set of sine waves, and any shape wave can be created by combining a set of sine waves. The Julia sets are akin to a Fourier analysis of the Mandelbrot set.

QED is likewise couched in terms of the addition and multiplication of complex numbers, and the waveforms of matter are directly related to QED. Furthermore, the properties of matter are a reflection of the waveforms. A simple form, such as a sphere, gives helium its exotic properties, so we should not be surprised if sophisticated forms such as the Julia sets have quite unexpected properties.

Describing Form

The intricate forms of the internal wavefunction are reflected in the composite probability density of subsystems, the external form of composite entities. To understand this external form we need to add an infinitesimal amount of calculus to what we have already discussed. We have already encountered many curves in the form of waves. The calculus provides a way to measure curvature, and how the curvature is changing.

Basic calculus

We start with the concept of the 'slope of a line' such as the two on the graph. If the change along y is Δy when you change x by Δx, then the slope is defined as the ratio of the two. The graph of a car traveling at a constant

speed is a straight line when time is plotted against distance travelled, and the speed is the ratio of 'how many miles' to 'how many hours.'

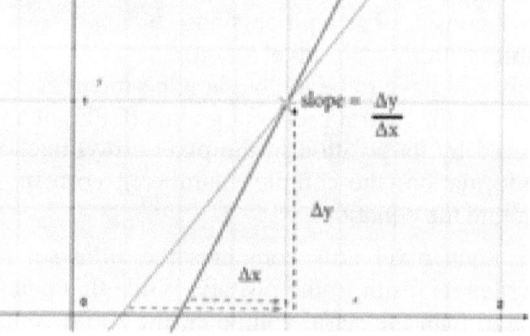

The red and green lines on the graphs have different slopes.

It can be seen by inspection along the x axis, that when the red changes x by 3, the y changes by 4. When the green changes x by 2, the y changes by 4. So the slope, rise or gradient, of the red line is 4/3 while that of the green is 4/2 or just 2.

This ratio remains constant if instead of moving the green four whole units along x, we move it just a small amount instead. What if we make the small

$$\text{slope} = \frac{\Delta y}{\Delta x} \qquad \text{speed} = \frac{\Delta miles}{\Delta time}$$

amount really, really tiny; shrink it as close to zero as we can imagine without ever actually reaching exactly zero. The ratio of these two 'infinitesimals' will always remain constant. This limit is the 'derivative' of the curve, and it can be used for curves as well as straight lines.

$$\frac{\Delta y}{\Delta x} = \frac{4}{2} = \frac{0.000004}{0.000002} \xrightarrow{\Delta x \to 0} \frac{dy}{dx}$$

If the red line is a distance-time graph of a car, and it travels 40 miles in 2 hours, we know that its speed is 20 mph. We observe the rather trivial point that the instantaneous speed (derivative) at 2 hours is also 20 mph, which is to be expected at constant speed.

Next we will look at graphs with curves in them, such as the magenta line which is y=x2. What can we say about its derivative, its instantaneous speed, at the point where x=1. The red line only approximates the slope, and it crosses the magenta in another place. If we rotate the red line counterclockwise a little, however, the two points merge into one point. The green line is the 'tangent' to the curve at that point. The slope of this line is the differential of the quadratic curve at the point where x=1.

It is clear that the tangent is different at each point, for instance it is a horizontal line when x=0 so the differential is also 0, and it has a negative slope when x is negative.

Now calculus has been around for a few centuries, and mathematicians do not go around measuring things, they have equations. A particularly useful one gives the derivative of any power of x. It gives the correct answer of 2 without measuring the graph.

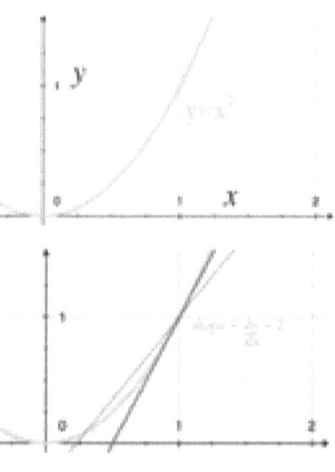

The derivative, or slope of the graph is itself a function of x. Whatever x is, the slope will be always be 2x. For example, at 0 where 2(0)=0, we have our horizontal line as the tangent. At x=100 the slope will be 200, while when x is negative, the slope will be negative, a downwards slope. We can graph the derivative against x and get the line y=2x.

$$\frac{dy}{dx}x^n = nx^{n-1} \qquad \frac{dy}{dx}x^2 = 2x^1$$

We can now ask about the rate of change in the derivative, the 'second derivative' of the magenta curve. This is the slope of the derivative graph, and we simply use the formula on the derivative to get 2, a constant. This is constant acceleration, and it has a derivative of zero, it is a horizontal line.

$$\frac{d^2y}{dx^2} = 2$$

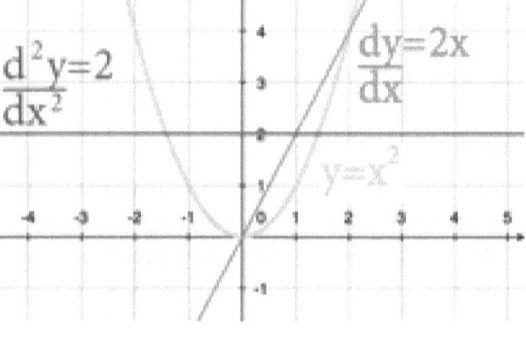

If the magenta line is a distance/time graph of a race car, it starts off at high speed as it enters the arena, and decelerates at a constant rate until it is zero at the royal box, where it turns around

$$y = x^2 \qquad\qquad \text{distance at time}$$

$$\frac{dy}{dx} = 2x \qquad\qquad \text{speed at time}$$

$$\frac{d(\frac{dy}{dx})}{dx} = \frac{d^2y}{dx^2} = 2 \qquad \text{acceleration at time}$$

and accelerates at a constant speed until it leaves the arena. The calculus notation for the second derivative, the acceleration, is similar to that of the first.

Just as addition and multiplication have their inverses in subtraction and division, so differentiation has its inverse called integration. For instance, if you are given the time and speed of a car, how do you calculate the distance traveled? You integrate the speed, v, which we have shown is the derivative of the distance function.

Integrals

Differentials give a measure of how a curve is changing; integrals give a measure of the area encompassed by a curve. We start with the simplest case, a constant curve where the x contribution is always one, x to the zero power, such as the green line on the graph which is y =2. What we want is the area under the graph from x=0 to x=6. For this simple graph, there is an elementary way and a calculus way to do it.

The elementary way is basic geometry, the area, A, is just A=xy=2×6=12. This is the integral of the function y=2 from 0 to 6. The calculus way is to note that the area swept out by the graph when x changes a little bit, Δx, as it does in the red rectangle, the area is just $y\Delta x$. And the total area is just the sum of all these little rectangles from 0 to 6. Being somewhat sloppy in the use of symbols by letting Δy stand for what y is at Δx, the sum of these little rectangles is also the total area under the 'curve'.

$$\sum_{0}^{6} y\Delta x = 12$$

Inverse Calculus

The simple geometry method will not do for finding the area under the curved blue graph, but the calculus method will. We let the small change in x, Δx, tend to zero and an infinitesimal width. As Δx, the width, tends to zero, the height of the rectangle is y to a greater and greater accuracy. So the area of the infinitesimal rectangle is ydx. The value of y changes with x at the rate, dy/dx, the first differential, so y is a function of (dy/dx). The total area is simply the sum of these infinitesimal rectangles symbolized by an elongated S, and this is the integral of the curve.

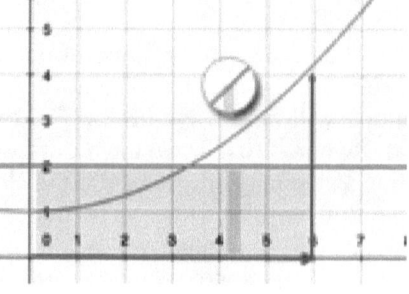

The two processes are interrelated such that if F has the differential D, the integral of D is F. Knowing that the differential of y=x2 is 2x, then we know that the integral of the line 2x is x2. We need go no further with integration as this is all we need to understand the discussion.

We are interested in the curvature of waves, such as sin(x), and we have already discussed the derivatives of the sine and cosine waves. The first derivative of the sine function is the cosine function.

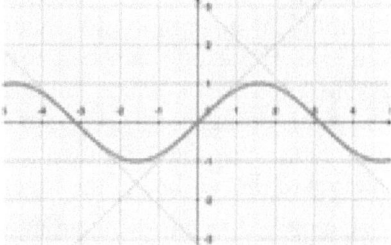

We can do the same for the cosine function, and the first differential of cos(x) is negative sin(x). This also means that the second derivative of the sine function is the negative sign function.

x	sin (x)	slope	cos (x)
0	0	+1	+1
180°	0	−1	−1
+90°	+1	0	0
−90°	−1	0	0

In plain terms, when the sine curve is at its maximum extent, its gradient is zero, and the change in the gradient is maximally in the opposite direction. When the sine is at its minimum extent, the gradient is maximal and does not change at all. Further derivatives reveal that they come in a cycle of four which repeats no matter how high a derivative you take.

x	cos (x)	slope	−sin (x)
0	+1	0	0
180°	−1	0	0
+90°	0	−1	−1
−90°	0	+1	+1

Furthermore, the minus sine and the plus and minus cosine are identical to the sine wave phase shifted by either a ¼ or ¾ of the period, ±π/2 (the cosines), or by a ½-period (the minus sine). Taking the derivative of one of these circular functions is equivalent to advancing its phase by ¼-period (90°, π/2 radians). This periodicity of four in the sine derivatives is reminiscent of multiplication by the rotation operator, i, which also has a cycle of four: +i, −1, −i and +1.

$$\frac{d\sin(x)}{dx} = \cos(x)$$

$$\frac{d^2\sin(x)}{dx^2} = -\sin(x)$$

$$\frac{d^3\sin(x)}{dx^3} = -\cos(x)$$

$$\frac{d^4\sin(x)}{dx^4} = \sin(x)$$

Everything about the shape of a sine wave is just a variant of the sign wave itself; it is a remarkably self-contained curve.

$$\cos x = \sin(x + 1\pi/2)$$
$$-\sin x = \sin(x + 2\pi/2)$$
$$-\cos x = \sin(x + 3\pi/2)$$
$$\sin x = \sin(x + 4\pi/2)$$

We will shortly be dealing with a very important

equation in quantum physics that is of the form shown on the right. Of course, it is not as simple as this, but given the above introduction to calculus, it should be clear that we solve this 'second degree, differential equation' by integration to obtain the first derivative

$$\frac{d^2y}{dx^2} = 2x$$

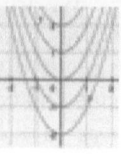

y=x2+k. The k is a constant to account for the fact that the graphs x2, x2±1, x2±2…. are a family of curves that all have the same derivatives. Going one step further, and knowing from tables that the derivative of x3/3 is x2, we know that the integral of this will be of the function itself, y=x3/3+k.

Compared to this differential equation, the only integral equation we are going to encounter is stating the simple fact of confinement within a curve.

$$\int_S y\,dx = 1 \qquad \int_{\sim S} y\,dx = 0$$

This simple equation states that the integral bounded by a curve, S, is 1, and that the integral outside the curve, ~S, is zero. Put simply, all of y is inside the curve S, while none of it is outside. We need to extend this simple calculus just a little to understand how science currently describes the form of the wave.

Calculus with Multiple Variables

The first step is to note that all the curves, or functions as they are more properly called, we have mentioned are functions of just one variable, e.g., f(x)=x2. It is but a small step to visualize functions and derivatives over many variables, such as f(x,y,z,t). The derivatives now involve slopes along more than one dimension, which if we needed to, would take us into the realm of partial differentiation, which fortunately, we don't. Integrals are no longer 2-D areas but 3-D volumes, and 4D and onwards, hyper-volumes.

We mentioned earlier that quantum scientists use the

$\Psi \equiv \Psi_s \equiv \Psi(p_x, a_x, p_y, a_y, p_z, a_z, p_t, a_t)$	Internal, 1s orbital
$\Psi^2 \equiv \Psi_s^2 \equiv \Psi^2(p_x^2, p_y^2, p_z^2, p_t^2)$	External, probability density

Greek letter ψ (psi) to stand for a quantum wavefunction, such as the 1s orbital. This is a quantum wave over the four complex dimensions, so the wavefunction is a function of eight variables, the wave at a point in internal space and time. As the subscripts get annoying, they are often collapsed into one symbol, or just left out altogether and implicit, as we shall do.

The Schrödinger Equation

In the discussion so far, we have used simple 1-D waves as examples. This can, however, only hint at a more sophisticated analysis as waves in multiple dimensions can be quite complex. The illustration is of a standing wave in two dimensions (such as a drum head) and waves in 3-D can be even more complex. We will meet some 3-D waves when we get to discussing orbitals.

With the little calculus we have discussed, the reader is hopefully not be intimidated by the expressions listed that are central to the quantum description of form and, in a general sense, the description of mind and body in Unified science.

The psi function, ψ, is the internal wave/mind, and the external form is ψ2, the composite density of the body. The 1st derivative gives the curve to the form, whilst the 2nd derivative gives the change in the curve. Higher derivatives are possible, but in practice the second is all that is needed. (This is true for cars as well. Speed is the 1st derivative, acceleration is the 2nd, and a 'jerk' is the name sometimes used for the 3rd if it is ever needed.)

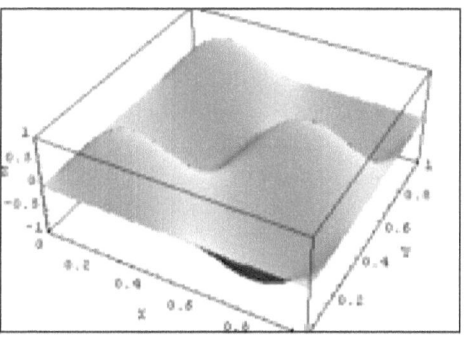

Ψ_s	&	Ψ_s^2	form of mind & body
$\dfrac{d\Psi_s}{ds}$	&	$\dfrac{d\Psi_s^2}{ds}$	curve to form
$\dfrac{d^2\Psi_s}{ds^2}$	&	$\dfrac{d^2\Psi_s}{ds^2}$	change of curve

Potential and kinetic

There is just one more thing before we reach the important equation, and that is potential and kinetic energy. This has nothing to do with the enormous energy tied up in the atomic nucleus, just the relationship between position and velocity in an interaction.

We are going to do a 'thought experiment' where you fall into a mine shaft that pierces through the center of the earth. We are ignoring any air resistance (it's a perfect vacuum) and any problem with heat at the center (the earth is stone cold). At the very start, all the energy is in the 'tension' of the relationship, the potential energy, P, and you are pulled very strongly towards the center. At this very start you have no velocity, and your kinetic energy is 0.

As you speed up and approach the center, the force of gravity diminishes and your acceleration decreases. At the very center, there is no force of gravity but you are moving at thousands of miles an hour. All the energy of the interaction is now in the released speed (kinetic energy) while the ten-

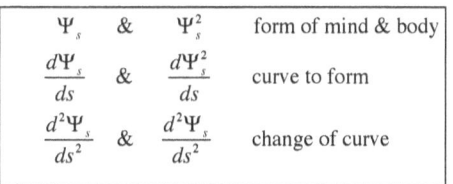

P=1, k=0
speed: zero
acceleration: max

P=0, K=1
speed: max
acceleration: 0

EARTH

P=1, k=0
speed: zero
deceleration: max

sion has disappeared entirely, the potential energy is zero. As you rapidly pass through the center, the force of gravity increases pulling you back to the center. You decelerate as kinetic energy is transformed into potential energy. Your head will briefly appear at the antipodes (speed is 0 and gravity is full strength) where all the energy of the interaction is potential again. The speed is zero at either end and maximum in the middle, while the potential energy is a maximum at either end. Yes, we are dealing with sines and cosines again. The kinetic energy follows a closed sine form, it had a zero node at either end. The potential energy, as we saw with the gluons, is an open cosine form with all the energy at the boundary. (More rigorously, we are dealing with (co)sine squared waves.)

This is a purely classical argument that ignores the internal, but as classical and quantum science do not disagree on such major topics, it suffices to allow the equation to be formulated.

You then repeat the fall in the opposite direction. Without any friction to draw away energy, the conservation of energy dictates that the sequence repeat endlessly.

In a similar way, we can treat the electron as if it is a ball interacting with the nucleus. All the energy of the interaction of a 1s orbital electron is potential energy when it is at the boundary node, and all the energy is kinetic when it is at the nucleus (which is transparent to electrons). The nucleus is also going to move in a complementary way, but being so massive it can be treated as unmoving without sacrificing much accuracy. The inertial mass, m, is just the reluctance of the electron to change its velocity (classical picture) or rotate its direction in spacetime (relativity).

At last we are ready to appreciate two very important equations of modern science, one an integral and the other a differential equation.

The first equation is simple and states that, if p2 is the probability density of the electron in the wave, ψ, the electron is 100% within the wave and not outside it.

$$\int_{\psi} p^2 = 1$$

Integer Solutions

The second equation is more sophisticated, it relates the kinetic/potential energy of the electron's wavefunction to the negative second derivative of the wavefunction (how the form of the wave decelerates, so to speak), along with the mass, the familiar constant π, and the quantum of action, h.

This is important because it is the scientific description in precise math language, of the mind of the simplest entity out of which the material world is constructed. It is the description of the mind of an atom, and is called the Schrödinger Equation. All of the following discussion will be based on this foundation.

$$-\frac{d^2\Psi}{ds^2} = \frac{8\pi m^2}{h^2}(K-P)\Psi$$

$$\Psi_n = \Psi_1, \Psi_2, \Psi_3 \ldots$$

Rather like the way that d2y/dx2 gave rise to a family of solutions, so too the Schrödinger Equation leads to a family of solutions that are numbered by the Principal Quantum Number, integer n. We have already encountered this number, it is the '1' in the name of the 1s orbital.

$$-\frac{d^2\Psi}{dr^2} = (\cos^2 r - \sin^2 r)\sin r$$

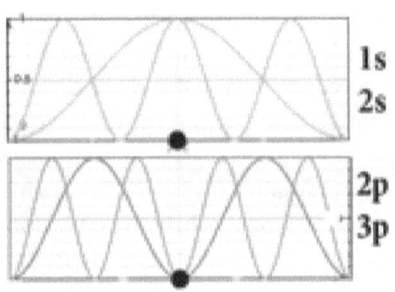

Without going into detail, we can change all the constants (such as mass, π and h) into units where they combine into 1. We can also set the free energy (this is the energy that changes during an interaction, the 'non-free' constant energy being the mass) as 1. We have seen that the share of potential and energy relate as the squares of a sine and cosine wave. Substituting all this into the equation, we get the much simpler expression that has an interesting graph. All the change in the form of the wavefunction occurs near the center, the boundary changes much less so.

Simple orbitals

This is only a simple introduction, but it should be clear how the equation of the wavefunction can be solved for a symmetrical ½-wavelength that fits with a maximum at the center and just one node, the boundary node.

To rephrase, current techniques are able to solve Schrödinger's Equation for the 1s orbital. This is the good news. The bad news is that we are unable to solve the equation for waves that have internal nodes (although approximations can be useful) such as the 2s (which is like the 1s but has an internal node) or the 2p which has an internal node at the very center.

	s				**p**				**d**		
	bound node	Internal Nodes			bound node	Internal Nodes			bound node	Internal Nodes	
		off center	on center			off center	on center			off center	on center
1s	1	0	0	*2p*	1	0	1	*3d*	1	0	2
2s	1	1	0	*3p*	1	1	1	*4d*	1	1	2
3s	1	2	0	*4p*	1	2	1	*5d*	1	2	2
4s	1	3	0	*5p*	1	3	1	*6d*	1	3	2
...	1	...	0	...	1	...	1	...	1	...	2

The 's' family of orbitals has no nodes at the nucleus, the 'p' family has 1 node there, the 'd' family has two there, and the 'f' family has three nodes at the center. The primary quantum number equals the total number of nodes in the orbital. These families of orbitals correspond to the blocks in the periodic table of the elements.

Such is the way that the electro-magnetic interaction structures the structure and function of everyday objects. The number of electrons in a neutral atom, however, is determined by the atomic nucleus, and this is the realm of the strong force.

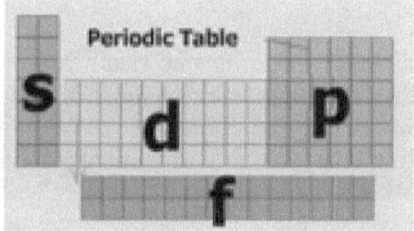

The Strong Interaction

As it does not make sense to consider an isolated quark or gluon, we have already discussed discussed the Strong force. The probability that a quark will absorb or emit a gluon is 1, it always does so. In comparison, the probability that an electron will absorb or emit a photon is much less, at ~1/137. This probability is the square of the internal probability amplitude, which is ~1/−11.7, and it is called α, the 'fine structure constant.' It is proportional to the square of the electric charge on an electron.

These probabilities are a relative measure of the strength of the interactions, so the strong force is considered 137-times more powerful than the electromagnetic force. The probability of a weak coupling is severely limited by distance, while gravity is intrinsically weak. The relative strengths are tabulated. Two of the interactions are long-range and two of them are short-range. We have seen that it is a balance between gravity and electromagnetism that is responsible for the structure of stars. At the other end of the scale, it is the balance between the electromagnetic and the strong that is responsible for the structure of the atomic nucleus in the stable and radioactive elements.

Coupling Constants			d
Strong	α_s	1	*short*
Electromagnetic	α	1/137	*long*
Weak	α_w	10^{-6}	*short*
Gravity	α_g	10^{-39}	*long*

The strong nuclear force that holds the atomic nucleus together—and the positive charge of the proton is always trying to get it away from any other protons—is a spillover from the strong force that binds the quarks together. A nucleon can exchange virtual pions, a quark and antiquark in a glueball. The virtual pions have a rest mass so they cannot exist for long or travel very far before disappearing before they amount to a pixel of the action. As the lifetime of the charged pions is 10 billionths of a second while the neutral pion's is only a trillionth-trillionth of a second, the charged pions have a much greater reach than the neutral pions.

Nucleons

Even for the charged pions, they cannot reach much more than the diameter of a proton. This is why the strong nuclear force is short range. The gluon shells merge somewhat and the two nucleons settle into a less-energy state by shedding real photons.

The two nucleons both have 'extra energy' problems. We have already seen that the extra pixel of color makes the neutron more massive than the proton. The close confinement of the positive charge in the proton also involves extra energy—it would take energy to force three ⅓ positive charges together, and this adds to the proton's mass, but not quite enough to make up the color energy.

The positive charge makes it impossible for the strong force to hold two protons together, the strong force makes them somewhat sticky, so they can oscillate for a very short time before the electromagnetic force drives the two apart.

The feeble neutral-pions are also unable to hold two neutrons together, the binding energy is too feeble and they fall apart.

If, however, in the brief sticky oscillation of two protons—a low probability—one of the protons does a reverse beta decay into a neutron—a weak process so also of low probability—a proton and neutron can exchange charged pions so that there is a resonance between the two. Both the problematic electrostatic energy of the proton and the problematic color energy of the neutron are reduced by being spread out over twice the surface. The energy saved by this resonance is more than the energy of neutron decay, and such a 'deuteron' is stable with the relatively small binding energy of 2.2 MeV.

This convergence of two events with low probability results in the very low probability of two protons fusing into a deuteron in the sun's core, and the resultant slow and steady conversion by the sun of nuclear energy into electromagnetic radiation. Even though the probability is almost infinitesimally small, the LLN compensates as there are an enormous number of protons in the million-degree core of our sun. There sufficient protons combine into deuterium so that 160,000 tons of gluon-energy is released as electromagnetic energy each second.

The deuterium nucleus picks up another proton in about a millionth of a second, and more energy is released as the problematic electrostatic and color energy spreads out over three nucleons. This diproton-neutron is a helium-3 nucleus. Over a longer stretch of time, and by various pathways, two of these combine into a four-nucleon helium-4 nucleus with two protons and two neutrons. So perfectly do the two pairs of fermions resonate in the nuclear wave, and so perfectly is the problematic energy spread out, that the helium-4 nucleus is so particularly stable that it is called an 'alpha particle' and is often shed by other 'radioactive' nuclei whose protons and neutrons are not well balanced.

The overall energy that powers all stars on the 'main sequence' is the conversion of four protons to one helium-4 nucleus, in which seven-tenths of 1 percent of the original mass is released as energy. We shall the creation of more complex nuclei, such as carbon and oxygen, when stars age and leave the main sequence in a later chapter.

The Weak interaction

The weak interaction, unlike the other three fundamental forces, is not system-building, it does create composites of simpler systems. Its role in the current era is the 'charged current' weak coupling involving the ±W-boson in the reverse-beta decay of protons and the beta-decay of neutrons. The weak 'neutral current' where neutrinos bounce off each other by coupling with uncharged Z-bosons has been experimentally observed.

To summarize, systems interact by coupling with their subsystems. The mutual change in the internal wave (which is immediate) and the external probability density (which follows over time) are the consequences of the interaction.

THERMAL RADIATION

To conclude this section on interaction, we will look at the simplest of interactions, two spheres of matter bouncing off each other. When two chemically-indifferent atoms (or collections of atoms) such as illustrated by the collision of two helium atoms, we saw the two bounce off each other rather like two billiard balls, the high-action state of the compressed area of the collision sending them both off in opposite directions. The collision is not perfectly elastic, not all the energy goes into the rebound, as the disturbed virtual photons can escape as real photons. The colliding helium convert some of their kinetic energy of movement into the energy of real 'thermal' photons.

The reverse situation can also happen: a helium atom 'surfing' the wave an incoming thermal photon can absorb the energy and the atom accelerates to a higher speed. In a gas of many helium atoms, we end up with a steady state where the loss of kinetic energy to photons in collisions is matched by the gain of energy from surfing the photons.

This bath of photons is called "thermal radiation" and it is always present unless the atoms are perfectly stationary. The amount and energy of these photons depends on the kinetic energy of the atoms, the 'temperature' of the gas.

In the steady state just mentioned, any atom moving faster than average will tend to lose more energy in collisions than it gains from absorbing the photons around. Conversely, slow atoms will tend to gain more energy from the photons than they loose from collisions. The steady state that is reached between the thermal radiation and the kinetic energy is called 'thermal equilibrium' and the thermal radiation depends solely on the temperature, not composition (only strictly true for a 'black-body' that absorbs and emits with equal facility).

$$\frac{1}{2}mv^2 = \frac{1}{2}MV^2$$

$$\frac{v^2}{V^2} = \frac{M^2}{m^2}$$

$$\frac{v}{V} = \frac{\sqrt{M}}{\sqrt{m}}$$

In a steady state of a mixed gas of helium atoms and heavier cousins, the average speed of the two will be different, but the average energy involved is the same. The kinetic energy, K, is equal to ½ the mass times the square of the velocity. So a helium atom accelerating to twice its speed has four time the kinetic energy than it started with. From this we can calculate the ratio of the speeds in situations of entities with different mass as proportional to the square-roots of their masses. In a mixed gas of chemically-indifferent helium atoms (He = 4 daltons) and oxygen molecules (O2=32 daltons) the helium atoms will be moving almost three times faster than the oxygen molecules. (As the speed of sound is governed by this velocity, speaking in a helium atmosphere creates a squeaky, high-pitched voice.)

The thermal velocity can also apply to single neutrons during their 11-minute lifetime. A high-energy neutron, such as those released from a fissioning uranium nucleus, is moving at high speed and the probability amplitude wave of the nucleus is very localized. The probability density of the neutron is very small and concentrated. The neutron passing through ordinary water encounters many hydrogen nuclei (protons) and in the collisions the energy is redistributed with the neutron probably losing energy and the protons gaining it. The neutron is slowed down by the 'moderator' to thermal speeds, and the neutron wave spreads out.

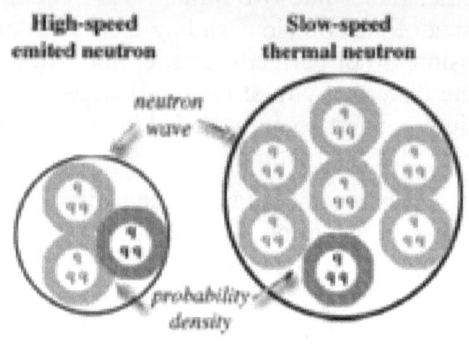

The mass of a neutron is about that of a hydrogen atom, and the best moderators are small atoms. An illustration of this is provided by the classical behavior of billiard balls. A fast ball colliding with a ping-pong ball will send it shooting away while its own progression is barely altered. A fast ball colliding with a stationary cannonball will barely shift it and just bounce of with hardly any alteration in its speed. But if a fast ball hits a stationary billiard ball, its kinetic energy is probably going to be shared about equally between the two.

After a few bounces in a moderator the neutron's speed is much reduced and the probability density of the thermal neutron becomes larger in size than an entire uranium nucleus of 238 nucleus. The 'cross-section' of the thermal neutron to be absorbed by the uranium nucleus is far greater than it was for the high-speed neutron. The absorption of the neutron creates such an imbalance that the uranium nucleus falls apart into two smaller nuclei and this fission releases a few neutrons that, in turn, are moderated and cause a chain-reaction of uranium fission. This is the basis for the controlled fusion in nuclear power plants.

At thermal equilibrium, the kinetic energy of the particles and the energy of the thermal photons is equal. There will be a spread of velocity and kinetic energy, naturally, but in any sizable quantity of gas there are so many collisions that the Law of Large Numbers dictates that this spread of is that of the calculated classical probability, the normal, or Bell curve.

The precise measure of deviation is called the 'standard deviation,' SD, and the Bell curve states that 99% of the speeds will be within ±3SD from the average speed, 95% will be within ±2SD, and almost 70% within just ±1SD of the average. The measure called 'temperature' is proportional to the average kinetic energy. The 'absolute scale' of temperature starts at absolute zero where the entities are motionless, and for each degree Kelvin,

K (which is ~C°+273) the temperature rises, the kinetic energy goes up by 'Boltzmann constant,' which is ~10-4 eV per degree K.

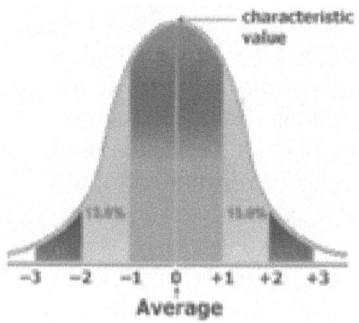

This is a classical calculation about the external aspects, and it works very well. So it was a great puzzle to scientists when it was found that the 'black-body spectrum' of photon energies did not have a 'normal curve' spread. It was skewed, as is shown in the diagram for four different temperatures, one of which is that of our sun. The high energy end of the curve was 'pushed in,' so to speak, and hardly any photon had an energy greater than 1SD of the average. It was this disparity that inspired the very start of the quantum revolution in science. This occurred when Max Planck hit upon the idea that photons were 'quantized,' that they each had just one unit of action. When he included this in his calculations he arrived at the skewed-spread of the thermal radiation. This epochal start of the 2nd scientific revolution is commemorated by Planck's Constant being the name given to one quantum of existence, and his name attached to all sorts of pixels.

This skewing by quantization is much more likely to end up in a low energy rather than a high energy quanta. The average energy of the photons is the same as in the collisions, but the characteristic energy is not the average energy of the photons.

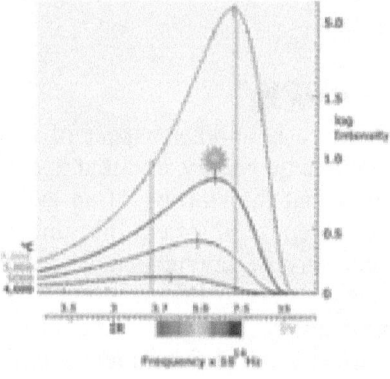

As the temperature rises, two things happen to the thermal radiation. The density of thermal photons increases (the intensity rises), and the characteristic frequency gets higher. This is the difference between the dull red and the bright white poker at low and high temperatures.

Using Boltzmann's constant, it is possible to graph the characteristic energy of photons all the way up to the extremes of temperature during the Big Bang, the theoretical limit being the Planck Temperature of 1032K and photons with an energy of 1028eV and a period of 10–44 sec, a period of one Planck time. From the graph it can be seen that at a temperature of 1014, the photons have the same energy, 1 GeV, as that locked up in the rest mass of a proton. At this temperature, there is more than enough energy to mangle spacetime into a proton-antiproton pair, and above this temperature there are as many protons

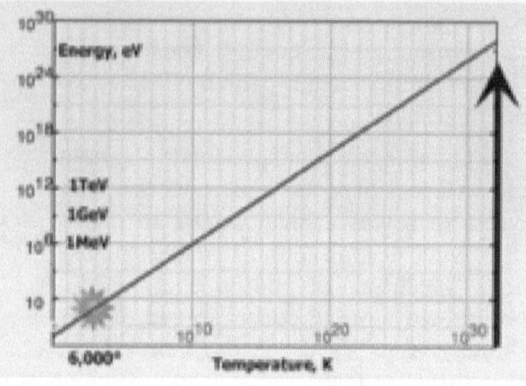

and antiprotons as there are photons, which at this temperature, are so numerous that their density exceeds that of water.

At even higher temperatures with densities approaching that of the atomic nucleus, the quarks, antiquarks and gluons are flying free and are also as numerous as the photons. As we are interested in falling temperatures, the point at which the temperature falls to where the energy is below the rest mass, and matter and antimatter annihilate faster than it is being created, is called the freezing out of, say, protons and neutrons. At energies much above their rest mass, all entities are moving at essentially the speed of light.

Entropy

There is another aspect of heat and temperature that is more abstract than kinetic energy, and that is entropy. The second law of thermodynamics states that the entropy of an isolated system of interacting entities can either be constant (at equilibrium) or must be increasing. This is just a technical way of saying that, absent an outside interference, the direction of change is always from a less probable state to a more probable state (the 'probability' being invoked is the regular, average kind, not the boson/fermion kind of fundamental entities). As there are a lot more disordered states than there are ordered states, entropy can also be described as the tendency to move from ordered to disordered states.

In the diagram, A is a system where all the kinetic energy is aligned in one direction—the whole system is moving in that direction. When it hits the ground, this kinetic energy is transferred to the random movement of the atoms, and the block heats up. There is only 1 ordered state A, but there are many disordered B states, only 2 of which are shown. Getting them all

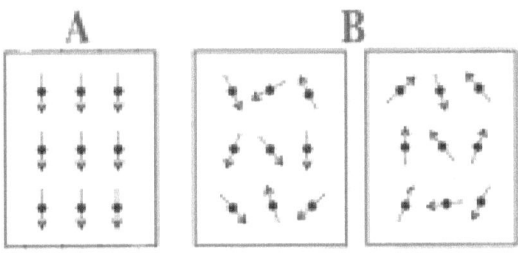

aligned again is going to take work, an input of energy.

As there are often a huge number of disordered states, the measure of entropy is related to the logarithm of the numbers involved. If there are, say, 1,096 random B states, then the entropy change, ΔS, in going from the single A state to the random B states, will be propor- $\Delta S \propto \log(1,096) = 7$ tional to 7 (as the already-encountered transcendental number e gives 1096 when multiplied together seven $e^7 = 1,096$ times, the numbers being chosen for convenience).

We have already seen that probability is as real as anything else is in modern physics. We have already seen that it is real enough to prevent the collapse of our sun when it ends its life as a black dwarf the size of the earth. So, although it is based on the rather abstract concept of probability, entropy is as real as energy and has to be taken into account. In chemical change driven by the principle of least action, for instance, both the energy, E, and the entropy, S, need to be taken into account in $H = E - ST$ understanding chemical transformation at a temperature of T Kelvins.

This combination is called the free energy, H, and it is this quantity that is important in chemical changes. While kinetic energy can explain why a gas expands into a vacuum, it is the entropy that insists on it

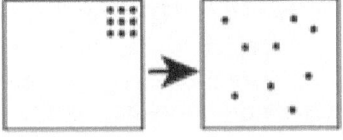

staying as a homogenous gas and not ever reassembling in one corner, leaving behind a vacuum.

Gravity will collapse a sufficiently large assemblage of atoms (as in the formation of stars) but then the probability is reversed, and the collapsed state is more probable than the extended states. In this case, the calculations involving the balance of energy and entropy are somewhat different.

PERFECT STORMS AND WAVES

The interactions of countless entities results in the waves that rule the current state of the environment. Sometimes these waves combine as bosons into a wave of unusual size.

This phenomenon of extreme constructive interference has only recently become the topic of intense study. Unfortunately, there are only two familiar examples, and both are destructive as far as humans are concerned, one in 2-D and one in 3-D.

Rogue wave

A 2-D example is when the waves on the ocean combine into a 'rogue wave' that is much larger, and much more destructive, than its component waves.

The wiki description is that rogue waves (also known as freak waves, monster waves, killer waves, extreme waves, and abnormal waves) are relatively large and spontaneous ocean surface waves that are a threat even to large ships and ocean liners. The rogue waves are not necessarily the biggest waves found at sea; they are, rather, surprisingly large waves for a given sea state.

Perfect Storm

A storm is a 3-D example, when the pressure waves driving weather, with a timecycle on the order of days, combine into a mighty and devastating storm. The experience of being at sea in such a storm is captured in the movie The Perfect Storm.

A "perfect storm" is an expression that describes an event where a rare combination of circumstances will aggravate a situation drastically. Since the release of the movie, the phrase has grown to mean any event where a situation is aggravated drastically by an exceptionally rare combination of circumstances. Perfect storm is nearly synonymous with "worst-case scenario" although the latter term carries more of a hypothetical connotation.

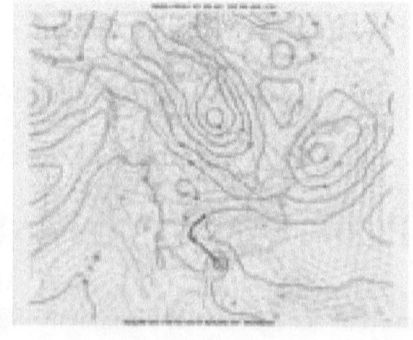

Goldilocks Wave

There are times in history when the environment is in the Goldilocks zone for systems to interact and become the subsystems of a more sophisticated system. The overall wave creates an environment which does not have too much or too little of what the interaction requires. The wave creates an environment that is 'just right' for system building to oc-cur—hence its name—the environment is an 'eden' in which the more so-phisticated system can emerge. A 'just right' goldilocks wave is synony-mous with "best-case scenario."

An example is the creation of 'primordial' helium in the helium-eden a few minutes after the hot big bang.

Before this time, the universe was a dense soup of free protons, neu-trons and electrons in a bath of gamma rays with energy greater than that of the binding energy of a deuteron. Any proton and neutron that did cling together would be quickly smashed apart by an incoming photon.

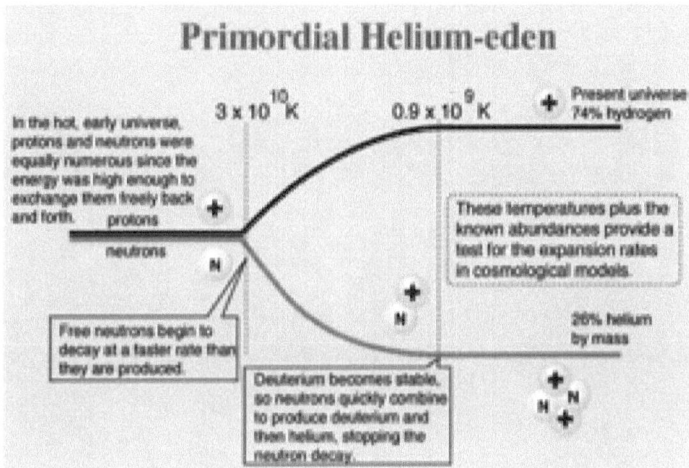

After this time, the thermal radiation could no longer prevent a deuteron from forming but the density had now fallen so low that protons and neu-trons hardly ever encountered each other.

During the short period of the goldilocks wave, however, the conditions were 'just right' for the helium-eden when all the neutrons ended up in helium-4 nuclei and those that did not get bound up decayed into more pro-tons and electrons. A similar goldilocks wave created an eden for atoms some million years later when the thermal radiation had fallen into the in-frared and the electromagnetic force could work to unite the protons and helium nuclei into atoms. The result was the 'dark age' when the Universe was 74% hydrogen, 26% helium in a bath of infrared photons. The eden for complex molecules took quite a few more billion years, a tale we will take up again when we return to discussing history and the Big Bang.

3.
EDEN OF
THE ELEMENTS

With the preceding overview of the fundamental entities out of which all things in the current physical universe are composed, we can proceed to an overview of how it all came to be. The history of the universe as it is currently understood.

This brings up the topic of the passage of real, positive time to make a history which is subtly different from the complex time of the p-metric (as witnessed by the difference between matter and antimatter which move in opposite directions in complex time while moving in the same direction in real time).

TIME AND HISTORY

One problem that arises in 'dating' the history of the first three minutes of the universe is that the temperature was so extremely hot and the average kinetic energy so extremely high that everything was moving at essentially the speed of light. The rule being that if the thermal energy is a few multiples of the rest mass of a particle, then the particles will be as abundant as the photons. We early noted that the boundary energy of open cosine wave of the real W and ±W pair gave them a rest mass of 90 and 160 billion eV, so if the thermal energy is one trillion eV, the real W-bosons will be as abundant as the photons, and all traveling at essentially light-speed. the weak force is on a par with the electromagnetic. The thermal energy during the very first moments was much greater than this, and the two forces are said to be a unified electro-weak force.

Even today, 'extreme cosmic rays' with an energy of a billion-trillion eV have been observed to strike the atmosphere, creating a shower of every type of boson and fermion-pair imaginable.

Absolute dating

Energies in the first few moments were much greater, implying that during the first few moments the internal wave of everything was moving essentially along the spatial axes and not at all along the time axis. It was almost a timeless moment for matter. Luckily, this state is normal for photons, so we can 'date' the first three minutes by the thermal spectrum of the photons, the blackbody radiation, and a history as it changes as the universe cools. The blackbody history also tells us what particles were in thermal equilibrium with the photons, and when they 'froze out' and could no longer be created by photons.

Most of the gamma ray photons generated by the hot Big Bang have yet to interact with anything; they have been flying unimpeded since they were released. These photons are all around in a vast abundance to this very day, 13 or so billion years later. By measuring their thermal spectrum we get an absolute measure of time since the moment of creation.

During all those billions of years, however, the energy of the photon has a gravitational effect, it has been attracting all the other quanta of energy (in bosons and fermions) in the universe. The universe, of course, has been tugging back during all that time. The photon has lost energy in opposing the expansion of the universe which has been slowed down. The gamma wave has lost energy and has been 'stretched' by the expansion of the universe into a microwave photon.

These 'relic' photons of the Big Bang—which outnumber the atoms in the universe by a hundred-billion to one—are the 'cosmic microwave background' radiation (CMB) that provide an absolute measure of time. For masses that have moved at substantial fractions of the speed of light in space since the 'decoupling' of radiation and matter, this blackbody radiation with a temperature just a few degrees above absolute zero will be skewed in predictable ways. Luckily, most things have only moved at tiny fractions of lightspeed since that epoch, and the time frame is barely shifted. So matter in the current universe is all on universal time. The CMB is the same no matter which direction you view the universe from earth, it takes sophisticated instruments to detect the slight shift caused by our rotation about the galaxy and its rush towards the Great Wall of galactic superclusters.

PRIMORDIAL ATOMS

At last we can return to the moment of creation. We saw that this involved the creation of an abstract Logos to control the internal wave. This operated on Nothing to create four orthogonal complex planes that were asymmetrically twisted apart into two sets of four axes, the p-metric with a left(−) twist and the s-metric with a right(+) twist.

The Logos operated on the p-metric, and the internal waves driven by the Principle of Least Action combined, changed and developed to generate the external history of the physical universe, the topic of which we will now discuss. The action of the Logos on the s-metric and the history of the substantial Spiritual realm will be taken up towards the end of this work.

Planck period

In the beginning, it was all Planck scale, and so to avoid a plethora of Plancks we will just abbreviate it as P. The p-metric was one P-length along each of the three imaginary spatial axes, and one P-time along the real time axis with an inherent chirality of left. It was at the P-temperature of about a trillion trillion trillion degrees, and had a P-energy of about ten thousand trillion trillion eV, about a microgram in mass units.

This 'inflaton' pixel of the 'false vacuum' as it is called had a negative pressure. Driven by this negative pressure, this pixel entered a period of inflation during which pixels doubled themselves in every tick of time and the energy was in all sorts of particles of every kind of charge. When the inflation started to wrench the color charges apart, the inflation was slowed as its outward rush was converted into the real energy of every kind of matter/antimatter pair and bosons. The universe was reheated to almost the P-temperature again by this energy dump into the universe which had expanded to solar-system size before the breaking kicked in.

TOE period

The kinetic energy was such that gravity was as powerful as the other forces, and this is called the TOE period when the forces were indistinguishable. Supermassive boson-fermions, called the X- and Y-bosons, with all axes a-jangle and an electrical charge of $\pm 4/3$ and a color charge, were in thermal equilibrium with the photons.

The now-slowed expansion caused the density and temperature of the universe to drop below the freezing-out point of these supermassive boson-fermions, and the matter and antimatter Xs and Ys decayed into electrons and quarks. The left-preference of the p-metric made this decay slightly asymmetric, and a tiny fraction of the $-4/3$ particle turned into an electron and a D-quark while the $+4/3$ particle turned into two U-quarks.

The temperature and density were so high that the thermal gluons and quarks were not confined and the three forces were on an equal footing.

Freezing-out forces

As the temperature and density continued to fall, the spatial limitation of the strong forces emerged and all the quarks became confined as protons, neutrons and others. This is the end of GUT period when the strong force froze out leaving the final two forces unified as the electroweak force.

Eventually the thermal energy fell below the mass of the W-bosons, and the weak force froze out leaving only the electromagnetic, and the four separate forces we recognize to this day. The neutrinos and antineutrinos that had been in thermal equilibrium, and in numbers equal to the photons, 'decoupled' from the other particles and went their separate ways.

The entities of composite quarks continued to be in thermal equilibrium, and matter and anti-matter were essentially equally present (there was that tiny imbalance from the X-boson decay).

Freezing out matter

Eventually, the thermal energy fell below that of protons and neutrons, and all the matter and antimatter nucleons annihilated leaving behind just the slight excess of matter. During the helium-eden period, some ended in the composite nucleus but most of them ended as free protons. The still-energetic photons were a hundred billion times as abundant.

Much later, the thermal energy fell below the electron/positron threshold and they annihilated leaving behind the electrons that exactly balanced the charges on the hydrogen and helium.

Much, much later when the gamma had stretched into the near infrared, the electrons combined with the protons and helium nuclei to create neutral atoms. The plasma froze into matter that only interacted feebly with the photons, and matter and radiation decoupled.

The atoms gravitationally aggregated while the photons went on to become the CMB; the neutrinos that had earlier decoupled from matter, and were as abundant as the photons, went on to be the best candidate yet for the 90% of the mass-energy in the universe, the quanta of the 'dark matter' that rules the large scale structure of the universe.

The initial pixel of spacetime in the inflationary phase doubled each Planck tick of time for a time period that is estimated to have lasted for 10–35 seconds which, while brief by human standards, is 109 quantum ticks. And the number of pixels doubled each tick. The final number of ticks is roughly $2^{10^9} \sim 10^{33,333,333}$. As the current volume of the visible universe has ~1030,000 cubic Planck units in it, so we are only able to see a fraction of the entire physical universe. The rest of the universe, though invisible, is at the same universal time.

STARS

The decoupling of matter and radiation happened about a million years after the big bang. The radiation had previously kept the plasma homogenous; once it lost its influence the atoms started to clump together by gravitational attraction.

The much more abundant dark matter, which had decoupled from radiation a million years earlier, had already clumped together into the scaffolding that attracted the atoms into the large-scale structure of visible matter that are the voids and strands of superclusters of galaxies that we see today.

In the following illustration of the large scale structure of the universe,[10] each pixel of color is a cluster of galaxies similar in size to our home galaxy, the Milky Way.

The relic neutrinos, just like the photons, had lost energy in opposing the universe's expansion, but unlike the photons they have a rest mass energy, which is constant and cannot be drained away. As we have seen, the rest mass is very small, only a few eVs, but there are a 100 billion of them for each atom, so their influence is considerable.

If their rest mass is as small as 0.1 eV, the mass energy of the relic neutrinos would be ten times that of the atoms, which is why these relic neutrino's are the best candidate for the Dark Matter that is ~10 times as great as that of the atoms in the universe. The nature of the dark matter must be considered to be still an open question as there are problems to be resolved.

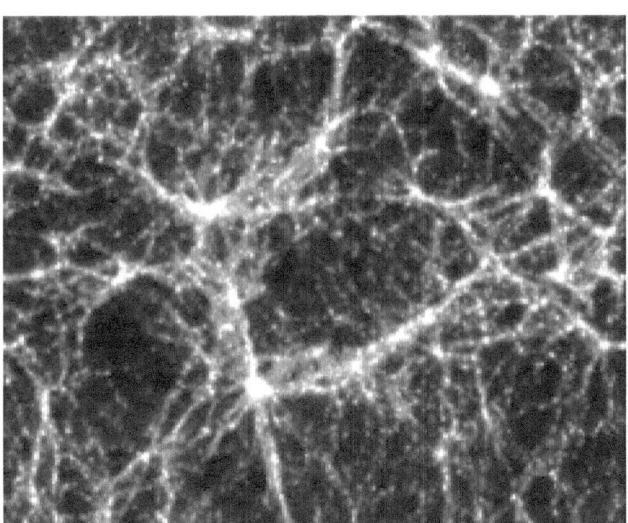

1st *Generation Stars*

It was only after this decoupling from the overwhelming number of photons that clouds of ~75% hydrogen-1 (protons) and 25% helium-4 atoms could start to condense around the dark matter.

The gravitational instabilities in these clouds fragmented them into galaxies and then into stars. The gravitational potential energy of the in-falling and colliding atoms was converted into kinetic energy, and the clouds started to warm up and emit thermal radiation. This radiation pressure opposed the gravitational collapse. Hydrogen and helium are not good radiation emitters, so the first generation of stars involved clouds that were ~100

10 http://www.antapex.org/large_scale_universe.htm

the sun's mass before gravity could impose as contraction over the poorly-radiated away electromagnetic heat that kept it extended.

As this collapse continued, the atoms re-ionized back into a plasma, and when the core temperature and pressure reached high enough values, the protons started to fuse together as a deuterium (hydrogen-2) nucleus (with the emission of a positron which combined with an electron into gamma rays) and so on to helium-4. The star ignited.

The radiation pressure outwards was now as great as the inward pull of gravity, and the star reached an equilibrium where the two were balanced. It took up a position as O class stars on the Main Sequence determined solely by its mass. As massive stars have a greater inward pull, they have to be at a higher temperature to reach equilibrium, and blaze with an intense violet-white light radiating 10,000 times as much energy as our sun does.

Before the stars ignited, the universe had been dark because all the gamma photons were now in the infra red, and this period is called the Dark Age. This period of darkness ended with the ignition of the first stars, and visible light reappeared in the universe. (It should be noted that all the billions of stars in the billions of galaxies, over 13 billion years of shining, have added but a tiny fraction to the number of photons in the CMB.)

Radiating so much energy, the 1st generation of stars ran out of hydrogen fuel in less than a million years as compared to the tens-of-billions of years for our G-class sun which is much more frugal with its initially-less fuel than such spendthrifts. When the hydrogen of the first generation of stars was gone the stars left the main sequence and started burning their helium.

The Main Sequence

The equilibrium between gravitational collapse and radiation inflation is reached at the same temperature for stars of the same mass powered by hydrogen-to-helium thermonuclear burning at the core. This relation of temperature and mass is called the Main Sequence. Our sun is about midway on this sequence, bracketed by massive stars radiating intensely in the UV-violet above, and low-mass stars dimly radiating in the red below.

A blue-white O-type star, with a mass 100 times that of our sun, has a surface temperature of 30,000K and emits 100,000 times as much energy as the sun does each second. It runs through its hydrogen in only ten thousand years before leaving the main sequence. A red M-type star with a mass just one-tenth of our sun, dribbles out just ten-thousandth of the sun's energy and can last over ten trillion years.

When the core of a star is so depleted of hydrogen that it can no longer create photons to oppose the inward gravitational pull, the core contracts and heats up as gravitational potential energy is converted into kinetic energy. The core heats up from the tens of millions of degrees in main sequence stars to hundreds of millions of degrees. The outer layers respond to the ferocious heat of the core by expanding and cooling. The aging start leaves the main sequence and the ultra-hot and dense core expands enormously the outer layers which cool off and the star becomes a red giant. The release of photons by helium burning now keeps the star expanded against gravity. Because the increased amount of energy is spread out over a larger area, each square centimeter will be cooler. The surface will have a red color because it is so cool and the surface will be much further from the center than during the main sequence.

Betelgeuse, the pink star in Orion, is an example of a red giant—actually a supergiant, as as it is much more massive than the sun and in its maturity it would have been an O-class star. Betelgeuse now has a size so great that its surface would almost engulf Jupiter's orbit. It is 150,000 as luminous as the sun, but much of this is in the infrared. This prodigious expenditure of energy is fueled by helium burning in the core, but exactly how far along it has progressed is currently unknown.

The energy released by helium burning liberates only a fraction of that released by proton fusion, and further stages release even less, so the lifetime is proportionally much less. Betelgeuse is expected to run out of all its fuel in the next million years, "any day now" by astronomical standards, and meet its spectacular end, to be shortly discussed.

Helium burning

Helium-4 will not undergo any change at the temperatures attained during hydrogen burning because it is so utterly stable. The quarks and gluons are fully-confined and in such a state of low energy that it takes what is called the "triple-coincidence" to get it to where system-building to carbon nuclei can occur. The development of this carbon-eden with a Goldilocks wave at the core of a star that is 'just right' for this to occur is directed by the Logos, the natural law that governs the development of the wavefunction.

A proton that tries to combine with the helium-4 to make what would be lithium-5 is quickly ejected after a brief stickiness, and the isotope has a half-life measured in fractions of a zeptosecond (10^{-23} s). Two helium atoms that attempted to merge into a beryllium-8 nucleus would vibrate in a strong stickiness but briefly and then fall apart into two helium-4s again. The half-life for the 'double-alpha' decay of 8Be is only 6×10^{-17} seconds.

When the core of a star contracts and becomes hotter and denser, helium nuclei are fusing together at a rate high enough to rival the rate at which their product, beryllium-8, decays back into two helium nuclei. This means that there are always a few beryllium-8 nuclei in the core, which can fuse with yet another helium nucleus to form carbon-12, which is stable

Ordinarily, the probability of this 'triple alpha process' would be extremely small. However, the beryllium-8 ground state has almost exactly the energy of two alpha particles and the kinetic energy with which the helium's collide at 100 million degrees is just right to put the beryllium-8 into an excited resonance that when combined with that of a third helium, 8Be + 4He, has almost exactly the energy of an excited state of 12C. These resonances greatly increase the probability that an incoming alpha particle will combine with beryllium-8 to form carbon. The existence of this resonance was predicted by Fred Hoyle before its actual observation, based on the physical necessity for it to exist, in order for carbon to be formed in stars. Experimental verification of these energy resonances gave very significant support to Hoyle's hypothesis of stellar nucleosynthesis, which posited that all chemical elements were formed from primordial hydrogen.

The final "coincidence" in the carbon eden of a star is that the Logos does not provide such a resonance for the carbon-12 to absorb another of the abundant helium and create an oxygen-16 nucleus leaving no carbon behind. Only a fraction turns into oxygen.

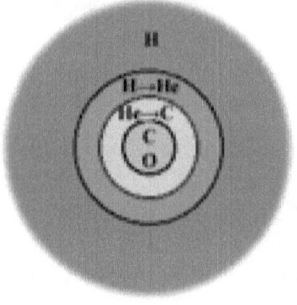

This is the final stage for a medium-sized star such as our sun. The core continues to shrink and heat up and the outer layers of hydrogen blow away as a planetary nebulae. But the core never reaches densities or temperatures sufficient for carbon and oxygen to start fusing together.

The star has a layered onion-like structure where the outer layers are still undergoing fusion. The star becomes a white dwarf as the hot innards are exposed, and once all the fuel is used up the star slowly cools to a black dwarf star. It is estimated that this takes a long time, and the universe is not yet old enough for such a white dwarf to cool off into the infrared.

As the star shrinks under the lash of gravity, another factor comes into play that prevents a total collapse, the fermion nature of the electrons in the star's plasma.

As mentioned earlier, the quantum probability of two fermions being in the same state is zero; it is impossible. As the volume of the white dwarf shrinks, the electrons reach a "degenerate" state in which they are on the verge of being forced to enter into the same state. The impossibility of this happening prevents any further collapse. As the white dwarf continues to cool, the volume remains a constant, held up by the 'degeneracy pressure' of the electrons, as it slowly cools through yellow heat to red heat and ends as a black dwarf. The mass is still that of the sun, but the volume is about the size of the earth's. The universe is too young for even the 1st generation of G type stars to have turned into black dwarfs.

Supernovae enrichment

When the core runs out of helium, it heats up to almost a billion degrees and the nuclei fuse into oxygen, then neon, sodium and magnesium, then silicon and phosphorus, and finally into iron and nickel.

For O-class stars such as the 1st generation of stars and our neighbor Betelgeuse, this is the end of the line, as iron has a minimum binding energy, and no more energy can be released by further fusion. There are no more photons to hold up the star against gravity and the core starts to collapse. The rest of the star has an onion layer structure, with the lighter elements around the core.

As the core fills up with iron, everything is happening so quickly that this final stage only lasts for minutes. The collapse continues and the core reaches a temperature such that the thermal energy is sufficient to create electron/positron pairs and reverse beta decay becomes possible and neutrinos are produced in abundance. These leave the core removing energy, and the core collapses accelerates.

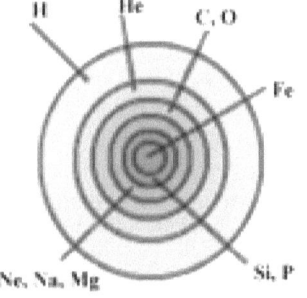

Bereft of support, the outer layers collapse and this sudden release of gravitational energy heats the entire star, and it explodes as a supernova.

In this maelstrom, a flood of neutrons are released and the thermal energy so great that some of the energy is absorbed in constructing atomic nuclei more massive than iron. The star shines out with the brilliance of 100 billion regular stars, and the elements heavier than helium (all called 'metals' by cosmologists) are scattered, enriching the primordial hydrogen and helium with all the other elements.

The force of this explosion so compresses the remaining core and smashes through the electron degeneracy pressure. Rather than be forced to do the impossible, the electrons take the only possible route and merge with the protons to create neutrons.

If the mass of the core remaining after the ejection is not too great (~2 solar masses), the neutrons, being fermions, also have a degeneracy pressure, and refuse to do the impossible and share a state. This can stop any further collapse, and the result is a neutron star that retains its size as it slowly cools off from the millions of degrees at formation as a gamma star through X-ray, UV, white, blue, yellow, red stages towards blackness.

The neutron star has the mass of the sun in a volume the size of a comet. The density of this 'neutronium' is that of the atomic nucleus. The neutrons are stable since it is the proton/electron pair, in such a situation, not the neutron that has the extra energy.

If the remaining mass is too great for the neutron degeneracy to resist the intense gravity, the collapse continues and a 'black hole' is the end result. To understand a black hole, we need the concept of escape velocity.

Black Holes

We have already discussed the interconversion of kinetic and potential energy in the simple pendulum in a gravitational field. A similar thing happens when a bullet is shot straight up in the air. As it rises in the gravitational field of the earth, it slows down until it comes to a momentary halt. All the kinetic energy of the bullet leaving the gun has been sapped away and stored as potential energy. It then falls towards the earth with increasing speed until (ignoring the frictional heat lost to the air) it hits the ground with its original speed. The energy is now all kinetic, and when it hits the ground this is randomized as heat.

The consequence of the gravitational disturbance of physical energy is attractive, and this applies to all forms of physical energy, including that of a photon. Just like a rising bullet, the gravitational field drains energy from a photon (a speeded-up version of how the big bang gammas became the microwaves of today). As a photon is always moving at lightspeed, it does not slow down, the energy is drained from the photon. It is redshifted as its wavelength is stretched, its period increases and its frequency decreases.

Gravity is, however, an extremely tiny effect. Two positive protons will, for instance, repel each other with a force that is a billion trillion, trillion, trillion times the attractive force of their mutual gravitation as a consequence of their physical energy.

For example, the earth and a human body both have equal numbers of positive and negative charges, and the virtual photons are all well-confined. The electromagnetic interaction between the two is minimal, and even a minuscule imbalance is quickly restored by a flow of current, such as a static spark or lighting strike.

Unlike electromagnetism, however, which has an equally powerful repulsion between like charges, physical energy is only attractive, and does not have an expansive effect. (We will encounter an expansive form of gravity when we get to the section that explains why we are using the term 'physical energy' rather than just plain energy.)

Gravitation is cumulative, and even an infinitesimal value can amount to a large value if it is multiplied by a big enough number. Most of the mass-energy of an atom is in the blaze of gluons that is the nucleons, and it is this energy en masse which gives rise to the gravitational mass of a composite body. The human body is a unified wave confining and giving form to ~30,000,000,000,000,000,000,000,000,000 quarks, electrons and an even greater number of gluons and photons. The earth has 2×1022 as many of them confined. Each one of these quanta of energy is gravitationally attracted to all the others, and the consequence is the force of gravity that keeps us on the earth's surface.

Sir Isaac Newton, while entertaining no hypothesis about the internal workings of gravity, derived these very useful approximations about the external nature of gravity that are still most useful to this day:

The mass of an extended rigid body can be treated as localized at a point within the rigid body, its 'center of mass' which, for an isotropic sphere is coincident with its geometric center.

The attractive force between two extended bodies is proportional to the product of their masses.

The attractive force between two extended bodies is inversely proportional to the square of the distance between their centers of mass.

The surface of the earth, and thus the position of the center of mass of the human body, is ~6,400 km from the center of the earth. The tiny force of attraction between all those zillions of quanta sums up to our 'weight' that holds us to the earth's surface, a force of 9.8 newtons for each kilo of mass.

In an airplane flying at 10 km (33,000 ft), we are at a greater distance from the earth's center, and the force is proportionately reduced. As the the proportional change is very small, however, and a person of 170 lbs, only weigh 8 ounces less than on the ground.

$$9.8 \times \frac{(6,400)^2}{(6,410)^2} = 9.77$$

Even on the International Space Station, at an altitude of 370 km, the force is still 8.76 and I would weigh 152 lbs. It is by no means a gravity-free environment. The weightless experience is a product of 'free-fall' and is akin to the weightless feeling momentarily experienced on a roller coaster.

$$9.8 \times \frac{(6,400)^2}{(6,770)^2} = 8.76$$

At the distance of the Moon, the force of attraction is reduced to just 0.003 newtons/kilo.

$$9.8 \times \frac{(6,400)^2}{(376,400)^2} = 0.003$$

The Moon is 1/80th as massive as the earth, but the surface is just ¼ of the distance to the center, so the surface gravity of the moon is 1.62 newtons, about 1/6th that of earth's, not 1/80th.

The sun is 332,950 times the mass of the earth, but its visible surface is 109 times as far from the center, so the surface gravity of the sun is just 28 times that of the earth's. Just as a bullet fired from the earth loses kinetic energy rising against the pull of the earth, so a photon leaving the sun's surface loses energy and is red-shifted, but only by about one millionth, and the red shift on leaving the earth is proportionately even smaller.

A bullet fired from the earth's surface is slowed until it stops, and then starts to fall back. The faster the bullet rises, the higher it will climb. But as the bullet rises, the force of gravity falls off. At a high enough initial velocity, the bullet will rise so far that the gravity is too weak to stop it so it never comes to a full stop but continues to climb. This initial velocity is called the 'escape velocity' (EV) and is a useful measure of the gravity gradient of a body.

The earth's EV is 11 km/sec, while a bullet from an M16 rifle travels at 10 km/sec and would be almost fast enough to never fall back again if air resistance didn't sap much of its energy.

It would, however, easily escape the Moon as its EV is only 2.4 km/sec. At the sun's visible surface, the EV is 618 km/sec, but the concept can be applied to any distance from the center. At the 'surface' of a sphere as large as the earth's orbit, the escape velocity is down to 42 km/sec, so even though the bullet can escape from the moon, it cannot shake free of the sun.

body	EV (km/s)
Earth	11.2
Moon	2.4
Sun	618
White dwarf	5,200
Neutron star	100,000
Black hole	299,792

The surface of a white dwarf has an EV of ~5,000 km/sec, and a photon rising against this gravitational field experiences a significant redshift.

The surface of a neutron star has such an intense gravitational field that the EV is >100,000 km/sec, or ⅓ the speed of light. When a neutron star implodes, all the mass collapses through a surface, the event horizon, at which the EV is the speed of light. A photon attempting to escape from the event horizon has an infinite red shift that is indistinguishable from no photon at all.   Nothing, not even photons, can escape from the event horizon. It emits no light, hence its name.

Structure of a Black Hole

No one has ever seen inside an event horizon, even in a thought experiment, so we can only theorize. Some think that the mass collapses into the infinite density of a point, a singularity. Quantum physics, however, suggests something that does not involve an infinite quantity as it connects distance with energy. The closer two entities get, the higher is their kinetic energy and thus the temperature.

As the collapsing star breaks through the neutron degeneracy pressure, rather than do the impossible, they first dissolve into a quark-gluon plasma. As they are compressed ever closer, the temperature reaches the stage at which average energy is capable of creating electron/positron pairs and then nucleon/anti-nucleon pairs. At even higher temperatures, the electromagnetic and weak interactions unite as the kinetic energy soars above the rest mass of the weak bosons and so on back to the earliest stages.

These are the last stages of the Hot Big Bang, and as the temperature rises in the collapsing core, the stages of the Big Bang reappear in reverse order. The outward pressure of this recreated primordial plasma prevents further collapse. Just how many stages are necessary to stop the collapse depends on the in-falling mass. The very largest Black Holes >billion suns probably have to go all the way and recreate a few pixels of false vacuum at the very center whose enormous expansive pressure holds up the layers above.

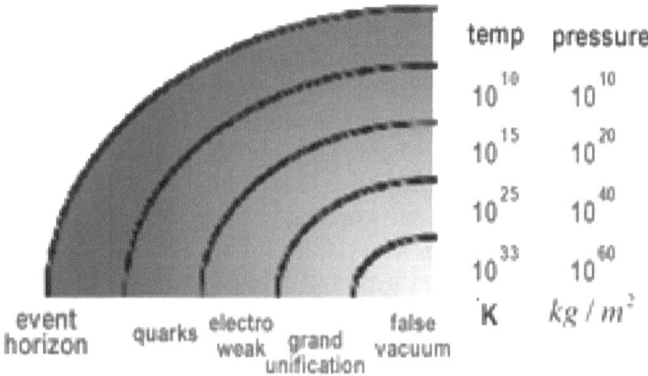

2nd and 3rd Generation stars

The first generation of massive stars within a few million years ran through their lifecycle ending up as neutron stars or black holes.

The black holes at the very centers of collapsing galactic-sized gas clouds merged into supermassive black holes with an enormous release of energy that it is thought to power the quasars that were commonplace in the first billion years of history and can be observed at the current limits of technology The formation of galaxies and their supermassive black hole centers is still an open question. As our galaxy has a billion-sun black hole at its center, we can suspect that the Milky Way home galaxy was a quasar in its early youth.

The now-enriched clouds of hydrogen/helium collapsed rather more quickly into the somewhat smaller 2nd generation of stars. Many of these are still around in the galactic halo of type 2 stars. Those that were large enough went supernova and further enriched the interstellar medium.

This metal-enriched H/He in turn collapsed into a 3rd generation of stars that are the type 1 stars (the somewhat inverse naming reflecting their order of discovery) of which our sun is a member.

Our galaxy went through its early stages until everything settled down ~10 billion years ago. The black hole that lives in the center of our galaxy is named Sagittarius A* (pronounced "Sagittarius A-star"). It is ~26,000 light-years from Earth and its event horizon is measured to be about 14 million miles across. This black hole would fit inside the orbit of Mercury and is estimated to have the mass of ~4 billion Suns.

The rest of the 100 billion stars in our Home Galaxy, the Milky Way, rotate roughly in a plane about this central point with a period of ~250 million years. If we call this a galactic year, our earth which is ~4 billion earth-rotation years old, our home is just 16 years old on the appropriate timescale.

All of the processes just described—from galaxy formation to 'metal' production in stars—occur under the direction of the Logos, and the end result was that the clouds out of which the 2nd generation of stars formed had a smattering of the 'metals' in them. The 'metals' are much more efficient at coupling with photons, and the 2nd generation stars tended to have smaller masses than the 1st generation. They were still large, however, and by about 5 billion years ago, they had also gone supernova and further enriched the interstellar medium.

While only about 1%, there were enough 'metals' around that when the 3rd generation of stars ignited and blew away the outer hydrogen and helium, there was enough of them around to form dust, which amalgamated into planetesimals and then into planets. The 3rd generation of stars to emerge were enriched with metals. Our sun is a 3rd generation star and has a significant enrichment in the 'metals' at 71% hydrogen, 27.1% helium, 1.3% carbon and oxygen, and less than 0.7% all the other 'metals'.

In the terms we have been using, the external development of the physical universe under the internal rule of the Logos created a Goldilocks wave that was a set of edens in which all the elements of our world were created. The stars were a set of 'wombs' in which the elements took their characteristic forms that were reflections of the abstract forms in the Logos; their forms and functions were 'inherited' from the Logos.

Here we will pause the progress of history of the universe and discuss some of the elements out of which everyday matter is con-

Element	Abundance (percentage of total number of atoms)	Abundance (percentage of total mass)
Hydrogen	91.2	71.0
Helium	8.7	27.1
Oxygen	0.078	0.97
Carbon	0.043	0.40
Nitrogen	0.0088	0.096
Silicon	0.0045	0.099
Magnesium	0.0038	0.076
Neon	0.0035	0.058
Iron	0.0030	0.14
Sulfur	0.0015	0.040

structed. Once freed from their stellar wombs, the atomic nuclei interacted with electrons to create the neutral atoms which predominate outside of stars.

It should be noted that while our local environment in which the sun and earth emerged is enriched with the astronomer's metals, on a cosmological scale the amount of primordial hydrogen and helium that has been processed into them is as yet a small fraction of the total. I have seen somewhere an estimate that it will take the universe at least another 100 billion years before the primordial storehouse is seriously depleted.

As mentioned, the entire universe is the same age as our bit a space, so the shortage will be a global, not a local challenge. As what has been done once can be done twice, the technological challenge of that far distant age will be to repeat the Big Bang—there is an inexhaustible amount of Nothing to convert into something, after all—and create fresh universes to expand into.

Atomic Nuclei

The multi-nucleon elements that are created during the excesses of supernova are extremely diverse, but only those that were stable or with very long half-lives made it into our everyday world. The resonance of proton and neutron is essential if the neutron is not to beta decay, and the composition of all small nuclei is ~50/50. As the nuclei get larger, the long range electromagnetic repulsion starts to overwhelm the sort-range strong force and extra neutrons have to be added to shore up the cohesion of the strong force, but this means the neutrons do not have an equal proton-partner and their instability becomes important.

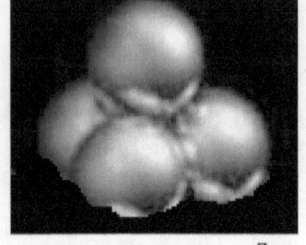

Eventually, no balance is possible, and all elements larger than Bismuth and Lead are unstable, they are radioactive.

The atomic number, N, is the number of protons—which determines the chemical character of the element—and the atomic weight, Z, is the sum of the protons and neutrons together, with the carbon nucleus with 6 protons and 6 neutrons $^{Z}_{N}X$ $^{12}_{6}C$ being given a mass of 12 daltons, where 1 dalton is ~900 MeV, the rest mass of hydrogen (the proton and the tiny contribution of 0.5 MeV by the electron).

Two nuclei with a different number of neutrons but the same number of protons (and thus the same number of electrons and chemical properties) are 'isotopes' of each other. Some elements have many stable isotopes—tin being the champion with ten in all—while others have only one, gold being an example. The stable elements can have radioactive isotopes such as a carbon where adding two extra neutrons to the usual six protons, six neutrons results in the carbon-14 isotope, and the neutron $^{12}_{6}C$ $^{14}_{6}C$ instability gives it a half-life of 5,730 years. At some point, a weak interaction between the quarks occurs, the neutron does a beta decay into a proton and the nucleus becomes a stable nitrogen-14 isotope.

The binding energy of helium is very large, and large nuclei sometimes behave as if some of the constituents are a helium nucleus interacting with the remainder. This 'alpha particle' is what leaves a nucleus in the alpha decay of large nuclei such as Uranium.

Radioactive Decay

The binding energy reaches its maximum at iron-56, to make larger nuclei than this, energy has to be added. Although all larger nuclei, such as gold, are theoretically unstable, their decay rate is so small as to be undetectable even over time periods greater than the age of the Universe.

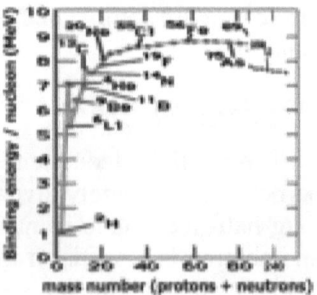

The most commonly observed decays of massive nuclei are:

Beta decay: When there is an excess of neutrons, their inherent instability comes into play. This excess energy is expelled by one of the D quarks as a virtual W– which decays into an electron and antineutrino which both leave the nucleon, leaving behind a proton. The fact that the electron does not carry away all the excess energy was the clue that led to the discovery of the otherwise unobtrusive neutrinos.

The Uranium-238 Decay Chain
Only main decays are shown
Gamma emitters are not indicated

Alpha decay: A helium 'alpha particle' can 'tunnel through the surface tension—a node in the wave—as the wave has a non-zero value outside. The probabilities involved are so very small, such as $10–44$ for an alpha particle reflecting off the surface barrier 1033 times a second as in a uranium atom with a half-life of a billion years.

Gamma decay: The nucleons are often left in an excited state by the departure or arrival of a neutron, an alpha or beta particle: this energy is shaken off as gamma rays. This leaves the nucleus in the ground state.

Loss of a helium 'alpha particle' leaves behind a nucleus with an atomic number minus two, and an atomic weight minus four. Loss of an electron 'beta particle' leaves behind a nucleus of atomic number plus one and unchanged atomic weight. Loss of a gamma results in an unchanged nucleus at the ground state.

Large nuclei, such as uranium, can split roughly into two when hit by a neutron—nuclear fission is a process successfully modeled by the 'liquid drop' model where the surface tension plays a major role.

In all cases of nuclear decay, the products have a greater balance between the strong, electromagnetic and weak decay. The process, such as in the decay of uranium-238, can have many steps until a stable balance between these forces is reached in a Lead or Bismuth nucleus.

			Per 100 atoms	
Element	*# p*	*# n*	# EC	# HB
H hydrogen-1	1	0	0.22	63
O oxygen-16	8	8	47	25.4
C carbon-12	6	6	0.19	9.5
N nitrogen-14	7	7	<0.1	1.4
Ca calcium-40	20	20	3.5	0.31
P phosphorous-32	15	17	<0.1	0.22
Cl chlorine-35	17	18	<0.1	0.03
K potassium-39	19	20	2.5	0.06
S sulfur-32	16	16	<0.1	0.05
Na sodium-23	11	12	2.5	0.03
Mg magnesium-24	12	12	2.2	0.01
All others			41.9	<0.01

The stable nuclei that are greater than just trace subsystems in living systems are just 11 in number, and the atomic weights and atomic numbers of their most common isotopes are here tabulated, along with their relative number in the earth's crust, EC, and a human being, HB.

All the atomic nuclei that are involved in the system-building interactions we will be discussing are stable. There are a few unstable nuclei around, such as carbon-14, and the energy released in the nuclear decay is almost always sufficient to disrupt whatever higher system it is a part of.

To summarize, the atomic nucleus is a system of interacting nucleons coupling with pions. The system wave firmly confines all but the virtual photons which escape to tangle with those of any nearby electron.

4.
ATOMS
AND CHEMISTRY

The atomic nuclei created in the star eden and released to enrich the primordial H/He attracts electrons and become neutral atoms. The electrons and nuclei interact by coupling with virtual photons, and they are all confined by a standing wave created by the resonance of all of their individual internal waves. The electron density and the density of coupling photons together have an external form that reflects the form of the internal wave, the system wave, which is directly determined by the Logos.

It becomes tedious to repeat phrases such as "the system wave is a composite standing wave resonance of the internal complex waveforms of the interacting subsystems and their coupling subsystems which is altered by subsystems entering or leaving the system" and "the overall external form of the system is the composite probability density of the interacting subsystems and their coupling subsystems confined by the standing system wave." The behavior and properties of a system are determined by its external form which is determined by the internal system wave, and its tendency to gain and lose subsystems in interaction is also determined by the internal system wave. We shall lump all these aspects together and refer to the complex wave as the 'internal character' of the system, and the composite real density of the interacting and coupling subsystems as the 'external form' of the system.

Helium

When the confinement of the subsystems is almost perfect, as it is in the neutral helium atom, the character of the system is utter inertness, and the external form is essentially that of the classical bit of massy matter, a tiny solid sphere that behaves like a classical billiard ball.

It is only at temperatures close to absolute zero that the character of helium shrugs off its classical mask and reveals its true 'odd-to-classical-eyes' nature. Each proton and neutron has three quarks with a spin of ½, and the sum of these is always a spin of ½, so they behave as fermions, not as bosons. The neutral helium-4 atom has six fermions in total—two protons, two neutrons and two electrons—and these can sum to an integer so the helium-4 atom has a boson character. When the thermal energy is small enough not to mask this tendency, all the helium atoms settle into the same state; they all are in the same internal wave and the liquid behaves as a single, unified system.

"Liquid helium behaves like a fluid without viscosity and with extremely high thermal conductivity. It appears to be a normal liquid, but will flow without friction past any surface, which allows it to continue to circulate over obstructions and through pores in containers which hold it, subject only to its own inertia. Since even gases have viscosity, superfluids have less resistance to shear than a gas does. Despite its lack of viscosity, the liquid still has surface tension, which allows it to rise up the sides of its containers without any normal frictional restrictions to flow. This allows the liquid to flow up the sides of containers, over the top, and down to the same level as the surface of the liquid inside the container, in a siphon effect."11

At high temperatures, such as those found in the Big Bang and in stars, the average kinetic energy is so high that electrons cannot unite with atomic nuclei; if they do so, they are quickly knocked off again. This is the state of matter called a 'plasma.' When things are sufficiently cool, enough electrons can form standing waves around the nucleus to cancel out its positive charge and create a neutral atom.

Having so much rest mass, the atomic nucleus only jitters over a fraction of the center of the wave. The nucleus and the light electrons both resonate in the same wave, but the spatial extent of the massive nucleus is 10-12 the volume of the electron cloud. For this reason, when discussing the structure of the atom it is customary to ignore the quivering of the central nucleus and treat it as an unmoving point. The focus is on the shape of the surrounding probability density of electrons. When an atomic wave bounces off another, however, the nucleus follows along, just like the electrons do.

If both the positive and negative charges are massive, both quiver in the atomic wave at the very center. If, for example, a tauon replaces an electron in an atom such as in deuterium, the mass of the tauon is equivalent to the mass of the deuterium nucleus, and it is also confined to the very center of the wave. It spends all its time inside the deuterium nucleus which is now electrically neutral. Two such tauonic deuterium atoms can approach, the nuclei can touch and fuse into the stable He-4 nucleus with the liberation of the binding energy. This would open the way to fusion power if the 'technical problem' of the very short lifespan of the tauon can be overcome.

---

11 http://en.wikipedia.org/wiki/Superfluid

ORBITALS

As fermions find it impossible to share the same state, each standing wave can hold just two electrons, one going clockwise, one going anti-clockwise. As mentioned, the standing waves of the electrons are classified by the number of nodes the wave has at the center where the nucleus is. There are the 1s, 2s, 3s... orbitals with no center nodes, the 2p, 3p... orbitals with one center node, the 3d, 4d... orbitals with two central nodes and the 4f... with three. The nodes are spatially arranged, giving at every level one s-orbital, three p-orbitals, and five d orbitals. The external shape of the probability density of these is illustrated—as promised, these 3-D shapes can be quite more complex than any 1-D standing waves might suggest (and are generated by simply adding and multiplying complex numbers together to get the internal form of which the external form is but a reflection).

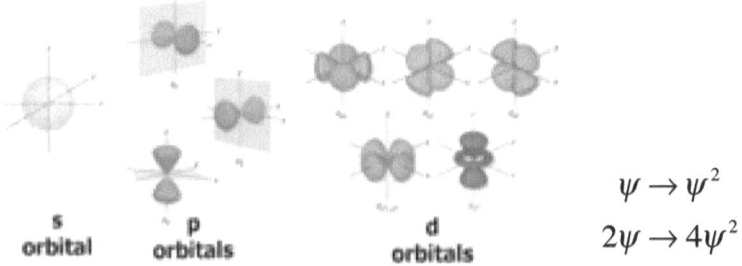

$$\psi \to \psi^2$$

$$2\psi \to 4\psi^2$$

s orbital   p orbitals   d orbitals

As we traverse the periodic table of elements in their neutral state, adding one proton at a time, each additional electron is added to either an orbital to make a pair, or to an empty orbital next highest in energy. Two electrons, while retaining their left-spin on the internal level, can share the 1s orbital by externally going clockwise and anticlockwise.

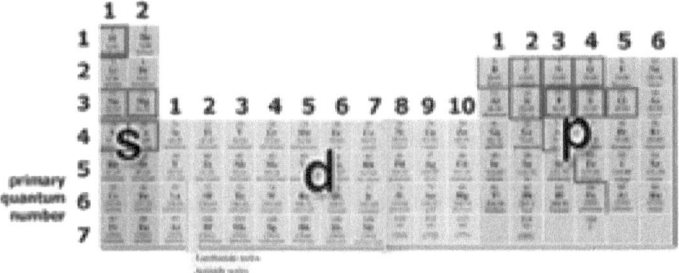

The internal waves, usually symbolized by the Greek letter ψ (psi), combine into a resonance with a probability that is four times greater, not twice, than that of the single wave, so the paired electrons are very stable.

Such paired electrons are in a low-action, high-probability state. Most of chemistry is driven by electrons seeking to enter into this paired state.

The single s orbital can hold 2 electrons, the triple p orbitals can hold 6, while the quintuple d orbitals can hold 10. It is this sequential filling of orbitals, and its repetition at each primary quantum number, that gives the familiar arrangement of the periodic table.

Chemical character

The character of the other 'noble gases', such as neon, argon, krypton, xenon and radon is almost as inert as helium with their subsystems almost as well confined by the system. These elements have their s-orbitals and three p-orbitals filled with pairs of electrons and are in a low-action state of stability.

In the other elements, such balance is not attained in their atoms, the wave does not perfectly confine the subsystems, and the Principle of Least Action drives the search for stability in the atom's interactions with other atoms. This is the 'chemical character' of the atom and this is inherited from the Logos via the form of the stable waves that result.

In fact, almost all of simple chemistry can be explained by a drive to attain a stable 'noble gas' configuration for the electron waves where everybody gets to have a complete set of electron pairs in the innermost, least energetic s- and p-orbitals.

Depending on the character of the atom, the internal confining wave, there are just three basic characters to atoms, the way they usually get their electrons all into the noble, stable, inert configuration:

1. Abandon electrons
2. Adopt electrons
3. Share electrons

The various elements have these abilities in various degrees. We will discuss the internal character and external form of a representative selection of the elements and their chemical character as we proceed.

HYDROGEN

We have already discussed how the internal traveling waves of an electron and a positron resonate together to form a standing wave that is called the 1s orbital. The equal-sized external probability density of the two creates the composite external form of a positronium atom.

To conserve momentum, the two particles coupling with connected virtual photons tend to occupy different halves of the sphere oscillating about the center at a frequency, that if a photon, would be in the ultraviolet.

Positronium is an unstable entity, but it can be made stable by replacing the positron with a proton, which behaves exactly like an overweight positron. The wave still confines the two, but while the electron density remains extended, the overweight proton density is now so localized that it is often considered a point on the atomic scale. The momentum is conserved as the electron and proton oscillate in the system wave of hydrogen.

The single electron of the hydrogen has the probability density that has the form of a 'singlet' electron in the ground state 1s orbital.

The wave has just one node, the boundary of the sphere, and the electron is confined within this boundary. Most of the virtual photons of the electron are connected to those of the proton, except for those that escape as the hydrogen atom, although overall neutral, has an electrical dipole from the charges tending to opposite sides. The external aspect of this sphere is a blend of the probability densities of the electron, proton and their coupling virtual photons.This is the state of lowest action, the ground state of the hydrogen atom.

A real photon with real energy can enter into this mix and kick the wave into an 'excited state' with two nodes, a boundary node and an internal node, and a greater spatial extent. This is called the 2s orbital. The external now has the form of two concentric spheres, the single wave has shells with a forbidden zone between them.

NOT TO SCALE

Excited states

Addition of more photons can create a wave with three nodes (one boundary and two internal) called the 3s, the 4s with four nodes and so on. The diagram illustrates the orbitals up to the 15s, although they can be much larger. It can be seen that they get closer together as they get bigger.

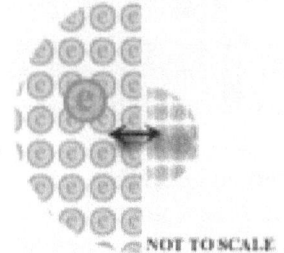

An electron in an orbital, say the 3s, can absorb a photon and add its twist, and the wave becomes a 4s wave. Another photon can add on making it a 5s wave. Precise experiments with lasers have put electrons into the ~100 level wave, a so-called excited 'Rydberg atom' that can be millimeters across.

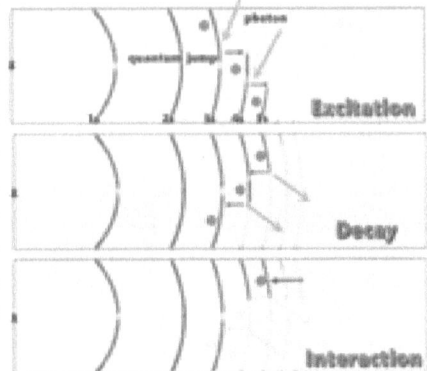

Such an excited state has a high

action and a low probability state, and the electron wave quickly spits out a photon and 'decays' to a lower energy state.

From the 5s orbital, it can emit a photon, and jump to the 4s orbital. With a series of jumps, the electron reaches the 1s orbital. The photons emitted are the characteristic 'emission spectrum' for that element. Hydrogen can also absorb these same photons 'that fit' and jump to an excited state, the 'absorption spectrum' of hydrogen, as illustrated.

EMISSION SPECTRUM OF HYDROGEN

ABSORPTION SPECTRUM OF HYDROGEN

The hydrogen atom is very lopsided, as can be seen from the crude diagram. Such a 'singlet' atom is highly chemically active as the hydrogen seeks a state of lower action.

Hydrogen chemistry

There are three ways in which the hydrogen can rid itself of the unbalanced, high-action single electron.

Accept an electron so the singlet becomes a pair, a negative hydride ion (H–). Only powerful electron donors, such as sodium, can force a hydrogen atom into this state with two electrons managed by one proton.

Share the singlet electron with another atom so that both end in a low-action state. The simplest case is two hydrogen atoms, sharing their electrons as a matched pair. This is almost like a helium atom with two separated positive centers and paired electrons filling the 1s orbital. The 1s atomic orbitals of each atom blend together into a 1s molecular orbital. The composite electron density is somewhat dumbbell-shaped. The pair of electrons the two protons are sharing is a 'single chemical bond.' It is not as stable as a helium atom, and the heat of a match will break it up into 'free radicals.'

Get rid of the electron and use some other atom's pair of electrons. Two hydrogen atoms can share their two electrons with oxygen's six to make a super-stable state of four pairs of electrons for oxygen and a pair of electrons for each hydrogen.

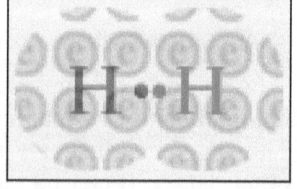

The hydrogen in a water molecule can abandon its electron and hook up with a pair of nonbonding electrons of an oxygen atom in another water molecule. The abandoned water molecule be-

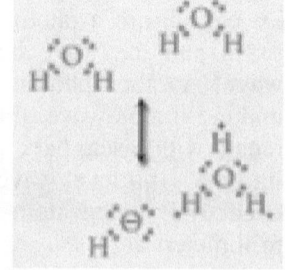

comes a negative hydroxyl ion and the three hydrogens now on the receiver resonate so they are identical, and the positive charge is spread out over all three hydrogens.

Most of the hydrogen outside of stars in the physical universe is in the form of neutral hydrogen molecules. In everyday life, however, the chemical character of the hydrogen atom in liquid water is of paramount importance, and it gives rise to the simple chemical concepts of 'acidic' and 'basic.'

Acids and Bases

The tendency of the hydrogen atom is to reach a balanced state by abandoning its electron and attaching to a molecule with a 'lone pair' of electrons to share. Such lone pairs are to be found in oxygen and nitrogen atoms, where some of the outer p-orbitals are filled with electron pairs. A water molecule can ionize; a hydrogen abandons its electron and fills its orbital with a lone pair of an oxygen of another water molecule.

This is usually written, ignoring the role of the second molecule, as: H2O ↔ H+ + OH–

In pure water, the concentration of ionized water molecules is 10–7—it has a pH of 7— and the product of the hydrogen and hydroxyl ions is 10–14. A chemist measures probability using the "Law of Mass Action" that, in the case of water ionization, states that the product of the concentration of hydrogen and hydroxyl ions will always be this 10–14.

$$\left[H^+\right]\left[OH^-\right] = 10^{-7} \times 10^{-7}$$
$$= 10^{-14}$$

The tendency for the hydrogen atom to abandon its electron and associate with a water molecule is much greater in the hydrogen chloride molecule, and a strong solution has a hydrogen ion concentration of 0.1 and a pH of 1. A solution of hydrogen chloride in water is a strong acid, H2O + HCl → H3O+ + Cl–. The concentration of the hydroxyl ion is proportionately mopped up by this excess of hydrogen ions, but their product is always 10–7.

$$\left[H^+\right]\left[OH^-\right] = 10^{-1} \times 10^{-13}$$
$$= 10^{-14}$$

The ammonia molecule, on the other hand, has a nitrogen lone pair that is very available. The hydrogen abandons its electron and attaches to the nitrogen atom. In strong ammonia, the hydroxyl concentration is high and the hydrogen ion concentration is a low 10–13, a pH of

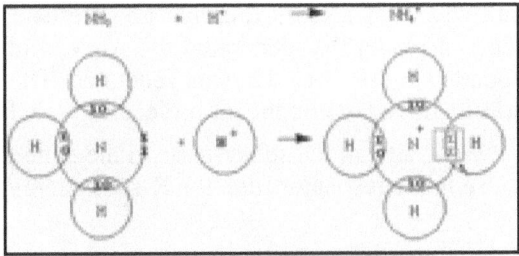

13. Their product is always a constant. $\left[H^+\right]\left[OH^-\right] = 10^{-13} \times 10^{-1}$
Ammonia solution is a strong base.

NH3 + H2O → NH4+ + $= 10^{-14}$
OH–

The ability to make electron pair bonds with other atoms is called 'valence' in chemistry, so hydrogen, which can make only one bond, is said to have a valence of 1. Chemists use a simple dash to represent a pair bond, so a hydrogen molecule can be written as H–H.

CHARACTER AND SUBSYSTEM

To summarize, an atom is a system composed of a set of interacting subsystems—the electrons and atomic nuclei—that are coupling with their subsystems—the virtual photons—all of which are confined and given external form by an internal atomic wave. In situations in which the confinement is not perfect, the atom can interact with other atoms by coupling with some of its subsystems, the electrons and virtual photons. If these interactions involve sharing a pair of electrons in a covalent bond, the atoms become the interacting subsystems of a higher system called a molecule.

A molecule is very similar to an atom (which can be considered a molecule of one atom), it is an array of subsystems, the atoms, interacting by coupling with their subsystems, the electrons and virtual photons. The interacting subsystems and their coupling subsystems are confined, and given form, by an internal molecular wave. If the confinement of the subsystems is not perfect, then some of the subsystems can be used as couplers in the interactions with other molecules.

The character of the molecule is inherited from the Logos, and a more sophisticated set of emergent properties manifest, one obvious example being to use entire atoms as couplers of interaction, such as in the hydrogen bond.

Systematic hierarchy

We see here the emergence of two classes of subsystems, each with a different character and role in the overall system. To illustrate this, we recall what was discussed earlier about music. There we also had two roles: the instrument that generated the wave, and the air as a resonator that responded to the wave that was generated. The instruments were massive and barely moved, while the air molecules were light and fast moving.

We shall call the subsystems in the generator role the G-subsystems and those in the resonator role, the R-subsystems.

As we have seen, both nuclei and and electrons contribute to the unified

SYSTEM	G SUBSYSTEMS R	
	WAVE GENERATOR	WAVE RESONATOR
	Few, massive, unmoving	*Many, lightweight, mobile, coupling*
Atom	Nuclei	Electrons

system wave of the atom. The nuclei are the G-subsystems—they are located at the center of the wave (as if they were generating all by themselves), and they are heavy and slow moving. The electrons are the R-subsystems—they are light and respond to the wave, giving it an external form, and it is the electrons that couple the interactions of atoms.

The wave generators in a symphony are digitally programmed as to what wave to create (the score). In this sense, the atomic nuclei are also digitally programmed, although permanently by their atomic number, as to what wave they generate.

The musical analogy breaks down when we come to the next way of classifying subsystems by their confinement by the wave. In other words, while the G-subsystems are almost always firmly confined, some of the R-subsystems are only loosely confined and can be used as coupling subsystems to interact with other systems.

While a system interacts with other systems by coupling with its subsystems, not all the subsystems participate. Atoms interact by coupling with electrons (and photons) but never couple with atomic nuclei (or pions, quarks and gluons). Atoms only couple with their outer electrons, the "valence" electrons, but they firmly confine the electrons in the filled inner orbitals.

For instance, a neutral sodium atom will interact with other atoms by coupling with its lone, barely confined electron in the 3s orbital, but even fluorine cannot get at the electrons in the filled, inner 1 and 2 'shells'. These are core subsystems, and remain constant, while the valence electrons are variable and available for coupling. The G subsystems are always core subsystems but, while only some of the R subsystems are core, some are potential couplers.

The progressive filling of higher-energy orbitals is the cause of the periodicity of chemical character in the atoms as we progress up the Periodic Table of the elements. An atom with a lone electron in the 2s orbital has an internal character that is similar to one with a single electron in the 3s orbital, while an atom with 5 electrons in the three 2p-orbitals, and sorely lacking an electron to make a balanced three pairs, has a similar character to an atom with 5 electrons in its 3p orbitals.

SYSTEM	G SUBSYSTEMS R	
	WAVE GENERATOR	WAVE RESONATOR
	Few, massive, unmoving	*Many, lightweight, mobile*
	CORE	COUPLER
Atom	Nucleus	Electrons Virtual photons
Molecule	Nuclei	Atoms Electrons Virtual photons

This is how the Logos directs system building at the level of atoms. As each new system emerges, a more sophisticated set of emergent properties are inherited from the Logos. The opposite process of system dissolution is also under the direction of the Logos.

Dissolution

The opposite of systems coming together as subsystems of a higher system is the dissolution of a system into its subsystems as a set of lower level systems. Whatever emergent properties the internal character of the higher system had inherited from the Logos disappear in the dissolution. For instance, the very basic emergent character of being substantial, essential to the experience of life, that is inherited from the Logos, and is considered a fundamental character in classical physics, completely disappears when an object is dropped into the sun and becomes a plasma of free electrons and nuclei.

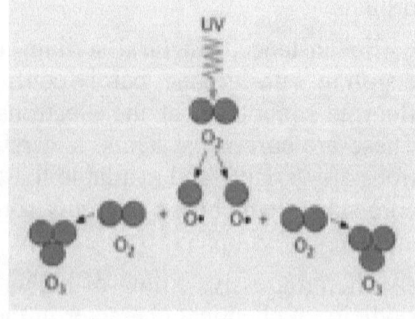

A photon of energy with just the right timecycle can split the system wave of an oxygen molecule into two atoms of oxygen, each with a singlet electron. In an excess of oxygen molecules, these free radicals convert an oxygen molecule into a triple-molecule with the wave and properties of ozone.

Another example of dissolution is electrolysis. If a current is passed through water, the ions go in opposite directions. The H+ is attracted to the negative electrode where it picks up an electron, pairs up with another H, and bubbles off as hydrogen gas. The OH– is attracted to the positive side where it gives up an electron, pairs up with another to create water and an O atom which pairs up and bubbles off as oxygen. Passed through molten salt, sodium metal appears at the negative while chlorine gas bubbles off the positive.

$$4H_2O \quad \to \quad 4H^+ + 4e^- \to 4H \to 2H_2 \qquad\qquad 2Na^+ + 2e^- \to 2Na$$
$$+ \quad 4OH^- - 4e^- \to 4OH \to 2H_2O + 2O \to O_2 \qquad 2Cl^- - 2e^- \to Cl_2$$

The process of breaking a polymer into its monomers occurs when a water molecule is added to a bond; it is hydrolyzed by enzyme activity. Proteins are hydrolyzed to amino acids, nucleic acids to nucleotides and polysaccharides to simple sugars. This is the process of digestion. The simple monomers are then reassembled into polymers or broken down for their free energy. A living system rarely repairs old components; they are usually broken down while a new component is made afresh.

The constant cycling of material means that almost all of a human body is replaced by new material every six months. The flow of subsystems in and out, however, makes very few changes to the system wave. Our bodies are constantly being broken down and built up; the mind and personality are not.

In the discussion we will focus more on system building and tend to neglect dissolution, but both are aspects of the workings of the Logos in the Physical realm. Much of the internal chemical character of the various atoms, and their capacity for system building, can be understood as a yearning, driven by the Principle of Least Action, to be a resonant wave that emulates that of the Noble Gases, so we start the discussion of the atoms of everyday life with these elements.

THE NOBLE GASES

The s- and p-orbitals are similar in energy level and in their space-filling shape about the nucleus. As mentioned, the filled 1s-orbital of helium firmly confines all its subsystems, and the helium atom does not couple with any of them—it has a valence of zero and is electrically neutral. The filled s-orbital is not so stable when there are empty p-orbitals close by in energy. But a very similar stability to helium is reached when both the s and p-orbitals are filled. These are the 'noble gases' that have little if any tendency to couple with their subsystems.

So stable is this arrangement that much of chemistry can be understood as being driven by the attempt of unbalanced electrons to reach a noble gas configuration, becoming a part of the unchanging core subsystems that are not used as couplers. A convenient shorthand in describing the electron in atoms is to show the 'noble gas' core with the outer electrons around it.

He	: $1s^2$
Ne	: [He] $2s^2 2p^6$
Ar	: [Ne] $3s^2 3p^6$
Kr	: [Ar] $4s^2 4p^6$
Xe	: [Kr] $5s^2 5p^6$
Rn	: [Xe] $6s^2 6p^6$

ALKALI METALS

The elements lithium, sodium, potassium, rubidium and cesium are all similar in having a singlet electron in an s-orbital. This is the valence R-subsystem electron about a noble-gas core. This electron is easily lost, leaving behind a sodium ion, a noble configuration but with a positive charge.

Li	= [He] $2s^1$	Li^+	= $[He]^+$
Na	= [Ne] $3s^1$	Na^+	= $[Ne]^+$
K	= [Ar] $4s^1$	K^+	= $[Ar]^+$
Rb	= [Kr] $5s^1$	Rb^+	= $[Kr]^+$
Cs	= [Xe] $6s^1$	Cs^+	= $[Xe]^+$

The alkali metals all have an electropositive valence of 1, they all have a tendency to give up the singlet electron becoming a positively-charged ion, this tendency increasing with atomic number. This is exemplified by the reaction of the alkali metals with water: Lithium generates bubbles of hydrogen; sodium reacts violently; potassium spontaneously ignites; while

SYSTEM	G SUBSYSTEMS R		
	WAVE GENERATOR	WAVE RESONATOR	
	CORE		COUPLER
Alkali metal	Atomic nucleus & inner electrons		s-electron

cesium explodes. This is the formation of sodium hydroxide (a strong base) from sodium and water. Na + H2O →
Na+ + H2 + OH–

As far as living systems are concerned, only sodium and potassium ions from the alkali metals are essential subsystems (while lithium does have medicinal uses).

These similar elements are all metals, a broad class of elements which are characterized by their loosely held outer electrons, the metals.

Metals

In cosmology, all the elements other than primordial hydrogen and helium are metals. In everyday life, and all other scientific disciplines, the metals are the many elements that, in macroscopic form, reflect light and conduct electricity.

		MP	BP			MP	BP
1	Hydrogen	-259	-253				
2	Helium	-272	-269	36	Krypton	-156	-152
3	Lithium	180	1317	47	Silver	962	2212
10	Neon	-248	-246	54	Xenon	-112	-107
11	Sodium	98	892	55	Cesium	28	690
18	Argon	-189	-186	74	Tungsten	3407	5927
19	Potassium	64	774	78	Platinum	1772	3827
26	Iron	1535	2750	79	Gold	1064	2940
29	Copper	1083	2595	80	Mercury	-39	357

When a number of atoms are together in a solid, the outer ½-filled orbitals merge together into a continuum of energy levels. The lower energy orbitals are being all filled with electron pairs—the valence band of orbitals—and the upper ones are empty—the conduction band. The atoms have a positive charge and are strongly cemented together by the sea of electrons. This gives many metals a high melting point—although mercury is an exception—and it takes a very high temperatures to turn them into a gas of free-flying atoms. The chart lists melting points (MPs) when the solid turns into a liquid, and boiling points (BPs) in centigrade when the liquid turns into a gas for the noble gas elements, the alkali metals, and some of the familiar metals.

It is the sea of electrons that gives metals their shine and ability to reflect light. The electrons are so mobile that the electromagnetic wave of an incoming photon sets them in motion, and like an antenna, sends up the exact same wave which radiates the same photon back. A photon of light striking a metal will cause the mobile conduction electrons to oscillate. This oscillation generates a photon that leaves the metal, i.e., the photon is

reflected by the metal. This is why untar-
nished, polished metals reflect light rays,
the principle behind the workings of a
mirror.

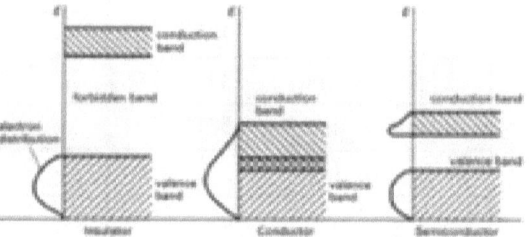

The phenomena is summed up in a
law of optics: the angle of incidence, i, equals the
angle of reflection, r.

The gap between the valence and conduction bands determines the mo-
bility of electrons in the solid, and its ability to conduct an electron current.
An 'electric current' is the bulk flow of electrons that have been kicked by
thermal energy into the conduction band.

If the gap between the two bands of orbitals is large, the solid is an insulator.

If the bands overlap, the valence electrons can move freely in the conduction band through the
solid—it is a conductor.

If the bands are close, a
few electrons are in the
conduction band and the
solid is a semi-conductor.

States of matter

This brings up the state
of atoms and molecules when there are a great many of them together, all
having the same average kinetic energy and the same temperature. Some of
the systems we will discuss have a strong interaction with each other, they
tend to stick together. Others hardly interact and fly free. Unless otherwise
stated, we will always be discussing systems at the standard temperature
and pressure, STP, which is roughly that experienced in everyday life.

Systems that are only barely sticky are gasses. An example is the hydrogen molecule which is in a helium-like self-satisfied state. Their kinetic
energy is more than sufficient to overcome any tendency to stick together,
and the molecules fly about freely, bashing into each other without harm,
and bouncing off the walls of any container (the gas pressure). The average
distance between the molecules is hundreds of times larger than the size of
the molecules, and gases are easily compressed. The kinetic theory of gases
uses statistical methods and classical probability to derive relationships
between temperature, T (the average kinetic energy), pressure P, (average
effect of bouncing off the walls), and container volume,V. The ideal gas
law (for molecules of zero size and zero stickiness) is that the ratio, PV/T,
is a constant. At absolute zero, 0 K, -273°C) the kinetic energy is zero, as is
P and V (which is clearly impossible for actual molecules).

Helium atoms are so indifferent to interaction that a temperature close to zero is needed before the not-quite-perfect cancellation of charge and the resultant tiny dipole is sticky enough to overcome the very small kinetic energy. At this point, the gas turns into a liquid, the boiling point of the molecule. This inherent stickiness of all molecules is called the van de Walls force, and it is at an absolute minimum in helium atoms.

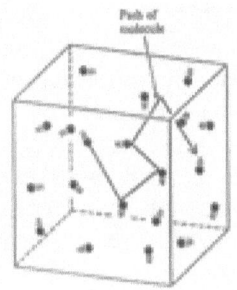

Helium-4 is a boson, it has an even number of ½-twist entities in its structure. At a low enough kinetic energy, the atoms all enter the same state—a giant molecule held together by one quantum wave. This state is quite unusual, with properties such as superfluidity, and quantized rotation. Helium-3, with an odd number, is a fermion, and its properties near absolute zero are even more bizarre.

A hydrogen molecule is not quite as balanced as the helium it is emulating, and they are somewhat more sticky. They have a higher boiling point (which is the same as the condensation point, just coming from the other direction). All the noble gasses have low boiling points, the stickiness increasing with the atomic number as the confinement by the wave becomes less and less perfect. Nitrogen and fluorine diatomic molecules also have tight confinement, and they have even lower boing points than monatomic argon.

At the other extreme, the element osmium, a dense cousin of platinum, is so sticky that it does not fly free as a gas until the temperature reaches 5027 °C.

He	-269 °C	Ar	-186 °C
H	-253 °C	O	-183 °C
Ne	-246 °C	Kr	-153 °C
N	-196 °C	Xe	-108 °C
F	-188 °C	Cl	-35 °C

At such temperatures, the average kinetic energy of the atoms and thermal photons is sufficient to knock electrons out of atoms, and above 10,000 °C, all atoms are ionized and we have a plasma of electrons and atomic nuclei, the so-called fourth state of matter. (The strange boson/fermion behavior of bulk liquid He-3 and He-4 probably merits a few more recognized states.)

Liquid and solids

When a gas condenses into a liquid, the kinetic energy is not sufficient to overcome the stickiness. The atoms are in contact, but otherwise are free to shuffle around. The are only two elements whose bulk state is a liquid at STP—bromine and mercury.

The atoms are pulled in all directions by the surrounding molecules, except at the boundary with something else, say air. Any imbalance at this boundary causes the phenomenon known as surface tension. For liquid mercury, the imbalance of forces at a glass-air boundary, the result called a meniscus, is convex shape, as is also seen with liquid water at a wax-air interface. In contrast, for liquid water at a glass-air interface, the imbalance of forces causes a concave meniscus.

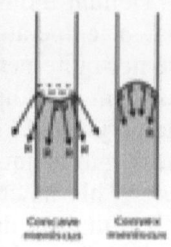

As the kinetic energy drops, the intermolecular forces become strong enough to hold the atoms and molecules firmly in place, the liquid freezes into a solid. All the elements become solids at a low enough temperature, except helium-3 which needs pressure to overcome its fermionic tendencies.

The melting points, where the transition from liquid to solid occurs, varies widely with the element, and some elements, like iodine, do not enter the intermediary liquid state but go straight from a solid to a gas, or vice versa. This is called 'sublimation,' and it also occurs in 'dry ice,' or frozen carbon dioxide.

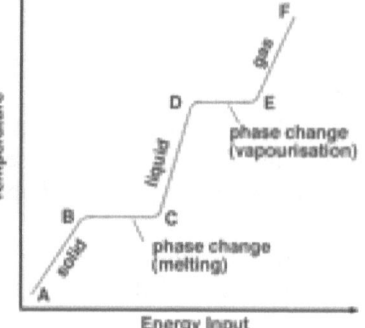

There is a binding energy when molecules change from the gaseous state to the liquid state, and from the liquid to the solid state. Added energy that does not go into kinetic energy, or raise the temperature but into the phase change is latent energy, and it has to be supplied from outside when melting or evaporating. Latent energy is a potential, not kinetic, form of energy. This is why the 'phase change' from one state to another occurs at a fixed temperature, the melting/freezing point, and the vaporization/condensation point. This potential energy is called the latent heat

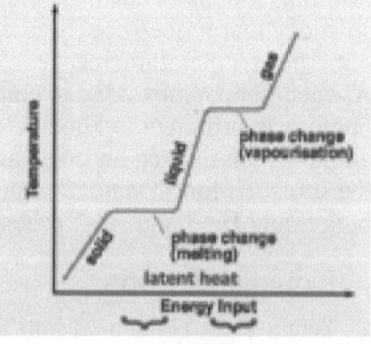

of melting and vaporization, and it is returned in full in the opposite direction as the latent heat of freezing and condensing.

At normal temperatures and pressures, the eighty or so stable elements in their pure form are mainly solids, most of which are metals. Just ten of them are gases and only two of them are liquids.

The solid state of bulk elements can come in different forms, called allotropes, and these can have a quite different internal character and external form.

Carbon, for example, has (at least) four allotropes: diamond (where the carbon atoms are bonded together in a tetrahedral lattice arrangement), graphite (where the carbon atoms are bonded together in sheets of a hexagonal lattice), graphene (single sheets of graphite), and fullerenes (where the carbon atoms are bonded together in spherical, tubular, or ellipsoidal formations). They have a wide variety of properties, from the black, slippery solid of graphite to the crystalline durability of diamond.

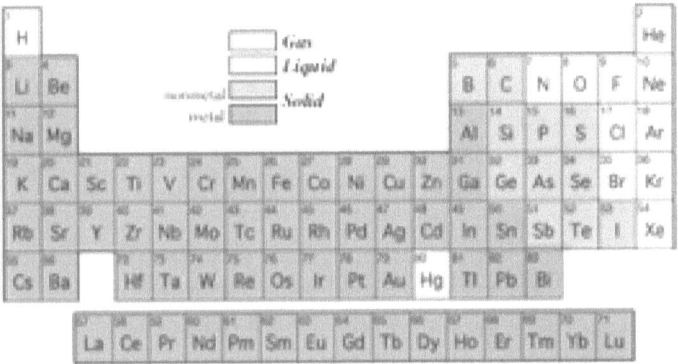

For some elements, allotropes have different molecular formulae which can persist in different phase—for example, two allotropes of oxygen (oxygen, O_2 and ozone, O_3), can both exist in the solid, liquid and gaseous states. Conversely, some elements do not maintain distinct allotropes in different phases—for example, phosphorus has numerous solid allotropes, all of which revert to the same P_4 form when melted to the liquid state.

As we continue to discuss the elements, we will be dealing with standard temperature and pressure unless otherwise indicated.

THE HALOGENS

Complementing the alkali metals are the halogen gases: they have one electron too few to complete a noble core of electron pairs, rather than the one too many of the alkali metals. The halogens are fluorine, chlorine, bromine and iodine. These are all avid acceptors of an electron to complete the noble shell, a configuration with a negative charge. The halogen atoms all have an electronegative valence of 1.

The tendency to gain an electron decreases with atomic number, a fluorine atom being so avid it can wrest an electron from almost any element, including xenon, and destroy any living thing. Iodine, on the other hand, is a mild and benign antiseptic.

F	= [He] $2s^2 2p^5$	F⁻	= [Ne]⁻
Cl	= [Ne] $3s^2 3p^5$	Cl⁻	= [Ar]⁻
Br	= [Ar] $4s^2 4p^5$	Br⁻	= [Kr]⁻
I	= [Kr] $5s^2 5p^5$	I⁻	= [Xe]⁻

Sodium chloride is the result when an electron slips from the sodium to the chlorine with the release of much free energy. The sodium ion is in the low energy neon configuration, and the chlorine is now in the argon state, which is much larger. Their opposite charges make them stick together by electronic attraction. They couple with virtual photons, and this 'ionic bond' results in high-

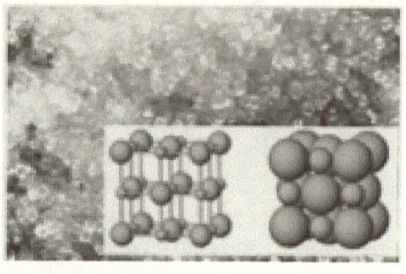

melting solids with a very regular, crystalline shape. The cubic shape of common salt crystals is a reflection of this atomic arrangement.

$$Na + Cl = Na{+}Cl{-}$$
$$[Ne]\ 3s1 + [Ne]\ 3s2\ 3p7 = [Ne]{+}\ [Ar]{-}$$

As far as living systems are concerned, only chlorine and a trace of iodine ions from the halogens are essential subsystems (while fluoride does have dental uses).

Alkali Earths

The properties of metals with two extra electrons in the outer s orbitals—such as calcium and magnesium—behave in a similar way to the alkali metals, just that they have a valence of 2. They are called the alkali earths. They are not so anxious to lose their

Be	= [He] $2s^2$	Be⁺⁺	= [He]⁺⁺
Mg	= [Ne] $3s^2$	Mg⁺⁺	= [Ne]⁺⁺
Ca	= [Ar] $4s^2$	Ca⁺⁺	= [Ar]⁺⁺
Sr	= [Kr] $5s^2$	Sr⁺⁺	= [Kr]⁺⁺

electrons, however, so they are milder in their properties than the alkali metals. (An object made of sodium would not survive the first rainstorm, while magnesium is metal in wide use.) They form very similar ionic solids and crystals to those of the alkali metals.

$$Ca + 2Cl = Ca2{+} + 2Cl{-}$$

The sold ionic 'salts' are held together by electrostatic attraction which gives them a high melting and boiling points.

	MP	BP
Sodium chloride	801	1413
Calcium chloride	772	1935

As far as living systems are concerned, only magnesium and calcium ions are essential subsystems.

OXYGEN FAMILY

The family of atoms that lack two electrons are not so alike as the halogens are. In water, the oxygen ion is unstable and quickly picks up a hydrogen to form a hydroxide ion. The sulphur atom is much less grasping and can take up a variety

O = [He] $2s^2\,2p^4$	O^{2-} = [Ne]$^{2-}$
S = [Ne] $3s^2\,3p^4$	S^{2-} = [Ar]$^{2-}$
Te = [Ar] $4s^2\,4p^4$	Te^{2-} = [Kr]$^{2-}$
Se = [Kr] $5s^2\,5p^4$	Se^{2-} = [Xe]$^{2-}$

of oxidized and reduced states depending on the circumstances. Only oxygen and sulphur are essential subsystems in living systems.

Oxygen needs two electrons to complete its noble-gas core, it has electronegative valence of two (as does sulphur). It can form ionic bonds with electropositive elements, such as calcium.

$$Ca\ +\ O\ \ =\ \ Ca2+ + O2-$$

Water

Oxygen can also form 'covalent' pair bonds in which each donate an electron pair to a molecular orbital that embraces many atoms, such as the three atoms in water, $2H + O = H2O$.

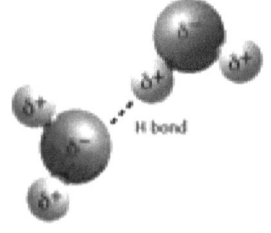

H bond

Water can couple not only with electrons and photons, it can also couple with hydrogen atoms as well, and it does so readily with other molecules, particularly other water molecules. The hydrogens switch allegiance from one O to the other, they oscillate back and forth. This H-bond makes the water molecule tend to stick together at preferred angles—the cause of the shapes of snowflakes. A molecular wave has a ground state, that gives form to the molecule, and excited states with extra nodes and waves that give form to the excited state of the molecule. Liquid water forms an intricate pattern of such 'hydrogen bonds' and this extra stickiness gives water anomalously high freezing and boiling points for its atomic weight.

Heavy molecules move slower than light ones so their phase transitions would be expected to occur at higher temperatures. This is true for molecules that only interact by the van der Waals force. Such molecules are incapable of hydrogen bonding; their hydrogens have no dipole and are firmly held. This inability to H-bond is a characteristic of the hydrocarbons, and the observation that, "Oil and water do not mix."

Light hydrocarbons have low temperature transitions, heavy ones have higher temperature transitions as illustrated by the boiling points of methane and ethane.

The ability to H-bond makes molecules very sticky, so they have unusually high transition temperatures for their molecular weight. This is seen in the boiling points of water and ammonia, with the oxygen having a greater H-bonding ability than nitrogen. It takes the hefty molecule heptane to match the boiling point of water.

molecule	weight	Boiling, °C	bond
He	4	−269°	Waals
CH_4	16	−160°	Waals
C_2H_6	30	−90°	Waals
NH_3	17	−33°	H-bond
C_7H_{16}	100	+98°	Waals
H_2O	18	+100°	H-bond
LiF	26	+1676°	ionic
Al	27	+2467°	metallic
Diamond	12	+4827°	covalent

This increase of phase transition temperature is not so great as that caused by ionic bonds, metallic bonds or covalent bonds.

The properties of water are of great significance for living systems, such as ourselves, in which the majority of the subsystems are water molecules. The internal character of the water molecule inherits qualities from the Logos that makes it the perfect resonator-subsystem that takes up the form of the composite wave from an array of generator-subsystems.

We have already encountered two types of systems with a small number of generator-subsystems creating a composite wave whose form is expressed in a large number of resonator-subsystems:

A symphony where an array of musical instruments create a composite wave of air pressure that expresses the form of the music.

A molecule where an array of atomic nuclei create a composite wave of electron standing waves that expresses the form of the molecule.

To these examples, we will add a third example. Living systems where an array of proteins create a composite wave of water structure that expresses the form of the living system.

As water-structure is to living systems as air-pressure is to a symphony, it behooves us to understand the character of the water molecule in more detail. For a start, the planet earth to which we are currently confined is said by astronomers to inhabit the Goldilocks Zone where it is neither too hot nor too cold but just right for bulk water to exist in all three states of solid, liquid and gas. The high latent heat of water as an unstructured gas is responsible for much of the variability in the weather, but we shall mainly focus on the structure of the liquid and solid states of water.

Structured water

We have described the interaction of hydrogen and oxygen as sharing a pair of electrons in a single covalent bond. It is not, however, an equal sharing as the electron wave is not symmetrical along the H-O axis, it is pulled asymmetrically towards the oxygen. The positive charge of the H-nucleus is only partially shielded, and the molecule is polar. There is a separation of charge. There is a positive side of the molecule where the two hydrogens stick out at an angle of 105°, and there is a negative side where the oxygen is.

It takes energy to overcome the electrostatic attraction between molecules. When the thermal energy is low enough, the angle between the two hydrogens dictates that the state of least energy is a hexagonal 'chair' that stacks up to create the open structure of ice crystals.

The covalent bond between hydrogen and carbon, unlike that with oxygen, has the electrons evenly shared and the 'hydrocarbon' molecules are not polarized. The thermal energy at −10°C is sufficient to disrupt the solid form of sluggish dodecane, a molecule of 12 carbons, 26 hydrogens and a mol. wt. of 170, while it takes another 10° rise in temperature to disrupt the ice-structure of water molecules with just a tenth the mass.

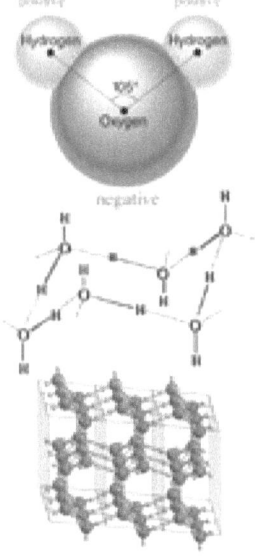

The ice melts into liquid water in which the molecules are mobile and the thermal energy is sufficient to break up any macro ice structure that attempts to reform on the microlevel. The liquid is more dense than the solid, and ice floats in water.

This is almost a unique property of water since most molecules form solids that are denser than the liquid form, and this is particularly important for life as it keeps the oceans from being almost all ice with only a thin layer of liquid water on top.

The water molecules in the liquid are strongly attracted to one another, and this gives water a strong surface tension at an interface with air. Air has no capacity to form hydrogen bonds, so the molecules at the surface are exposed to an unbalanced force being pulled into the liquid but not out by the air molecules.

The same thing happens with an oil and water interface—the surface tension of water squeezes out the oil and they separate. The oil is hydrophobic and the water will move such molecules so as to minimize the surface between them.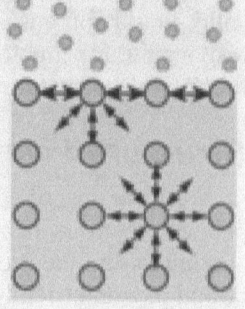

On the microlevel, a proportion of the water molecules in the liquid state will be in the preferred hexagonal form, and this proportion falls as the temperature and kinetic energy rises.

This holds for pure liquid water, but the proportion in the preferred form can be altered by other molecules. The simplest situation is when a molecule is soluble in water.

Solubility

While an ionic solid such as sodium chloride has a high melting point due to the attraction between the ions, they can reduce this by attracting water molecules and spreading their electric charge over a volume of water. The ions, now coated with a 'hydration shell' of water, fall apart and the salt is soluble in water.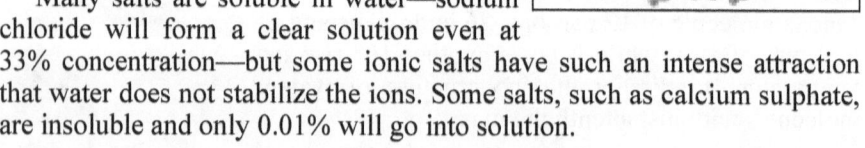

Many salts are soluble in water—sodium chloride will form a clear solution even at 33% concentration—but some ionic salts have such an intense attraction that water does not stabilize the ions. Some salts, such as calcium sulphate, are insoluble and only 0.01% will go into solution.

Molecules that can H-bond are also soluble in water. Examples are molecules with hydroxyl groups, such as alcohols and sugars, which are good at giving a hydrogen bond, and the amines that are good at receiving hydrogens.

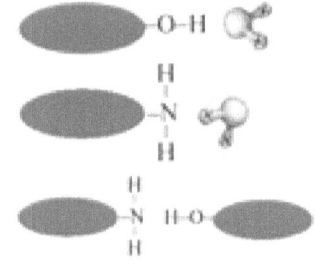

In some situations, no water is involved and the most stable state is a hydrogen bond directly between the nitrogen and oxygen, linking their attached molecules together in another level of system-building interactions directed by the Logos. Many of the emergent properties in the subsystems of life depend on such a balance of hydrogen bonding with water or with some other entity, even another part of a large covalent molecule.

Sulfur

Unlike oxygen, sulfur is a solid; it is not so avid as oxygen and tends more towards more equal bonds with itself and hydrogen. In the elemental form, the sulfur shares electrons with four others and they form a ring of eight bent into a crown. The molecular wave firmly confines the electrons, and solid sulfur is an excellent insulator that will not allow any current to flow even under a powerful voltage drop.

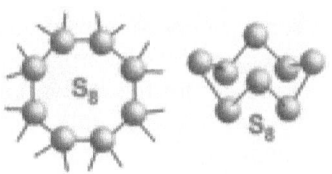

The counterpart of water is hydrogen sulphide, H_2S, which is a (noxious) gas as there are only weak hydrogen bonds to hold the molecules together. In bulk, it stays a gas down to $-60°C$ when it turns into a liquid that solidifies into crystals at $-80°C$. The liquid form has only a 20° range compared with the 100° range for water. While the sulfur equivalent of the hydroxyl radical, the sulfhydryl radical, is not as commonplace in living systems, the more covalent character of sulfur plays an essential role in many of life's subsystems.

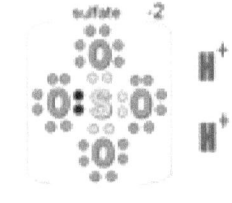

Sulfur will unite covalently with four oxygens to generate the powerfully-acid sulphate ion. While only sparingly used in living systems, the emergent properties of this molecule make it of preeminent importance to industry.

NITROGEN FAMILY

The family resemblance between atoms lacking three electrons in the outer p-orbital is even less than in the oxygen family.

Nitrogen and phosphorus tend to form covalent bonds, and with oxygen they unite to form the acidic nitrate and phosphate ions.

The atom of nitrogen is three electrons short of the balanced, low energy state of filled electron pairs. It can merge its sp3 valence orbitals with the 1s orbital of three hydrogen atoms to create the molecular wave of ammonia. It has a lone pair which readily shares with the H+ ion making ammonia a base and nitrogen a good complement to oxygen in hydrogen bonding capacity. Ammonia is almost as good as water in forming H-bonds, making it exceedingly soluble in water.

The ammonia wave has a resonance mode in which the N atomic nucleus 'tunnels' across the plane of the 3 hydrogens. This is an excellent

N	= [He] $2s^2 2p^3$	N^{3-}	= $[Ne]^{3-}$
P	= [Ne] $3s^2 3p^3$	P^{3-}	= $[Ar]^{3-}$
As	= [Ar] $4s^2 4p^3$	As^{3-}	= $[Kr]^{3-}$
Sb	= [Kr] $5s^2 5p^3$	Sb^{3-}	= $[Xe]^{3-}$

example of how the wave determines the particle's history, the massive nitrogen teleporting back and forth in the oscillating wave. The smeared-out densityy of the single N nucleus is, just like the 2p orbital, in two lobes of 50% with a node of 0 in the center. This is an example of entanglement over atomic distances of a relatively massive entity.

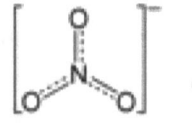 

We will see, on a larger scale, this ignoring of the spatial distance between particles when atoms jump about in a wave in the structure of macromolecules essential to life, the proteins and the nucleic acids.

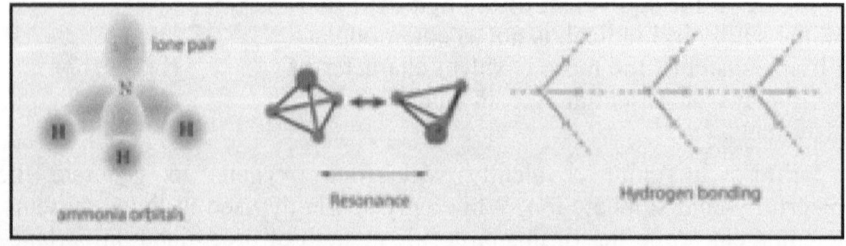

Both nitrogen and phosphorus are essential subsystems for life; the nitrogen with hydrogen in the ammonia-like state, and phosphorus with oxygen in the form of the phosphate ion. Of particular significance are the high-energy polyphosphate bonds whose making and breaking is the energy currency of life. The polyphosphate bond is high-energy compared to most of the other chemical bonds and breaking the phosphate bond can drive the formation of most bonds.

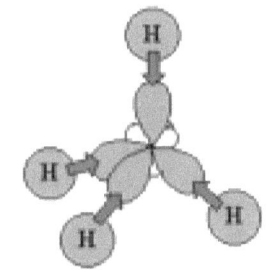

CARBON FAMILY

The family of elements with just two electrons in the p-orbitals are quite dissimilar in their internal character, as exemplified by the great difference between carbon and tin.

An isolated carbon atom is in a high-energy, asymmetrical state. It has an atomic number of 6 protons balanced by 6 electrons. If the orbital filling was simple, its electronic state should be:

$$C = [He]\, 2s^2\, 2p^2 \qquad N^{3-} = [Ne]^{3-}$$
$$Si = [Ne]\, 3s^2\, 3p^2 \qquad P^{3-} = [Ar]^{3-}$$
$$Ge = [Ar]\, 4s^2\, 4p^2 \qquad As^{3-} = [Kr]^{3-}$$
$$Sn = [Kr]\, 5s^2\, 5p^2 \qquad Sb^{3-} = [Xe]^{3-}$$

1s2 2s2 2p2

This, however, is highly-asymmetrical and the mismatched 2p electrons would have a very high energy. To balance things out, the s and p waves combine together to form four identical waves, called sp3 hybrid orbitals, with the form of lopsided lobes with one central node that point to the vertex of a tetrahedron. Each orbital has a singlet electron jittering about in it.

A hydrogen atom can merge its singlet 1s wave with a singlet sp3 to create a molecular wave filled with a balanced pair of electrons. Four such chemical bonds gives the molecular wave called methane. This system wave governs the density of five atomic nuclei, 10 electrons and innumerable virtual photons.

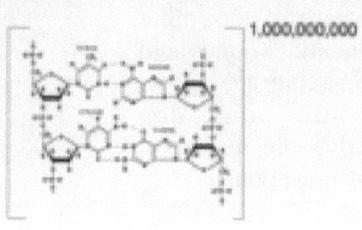

1,000,000,000

The containment is almost perfect and methane is not a very reactive molecule. The carbon atoms can link up in chains with a molecular wave that can span less than a dozen atoms, as in isopentane. The wave can be much larger, embracing billions of atoms as it does in the polymer called DNA.

Handedness

A carbon atom that is bonded to four different groups comes in a left-handed and a right-handed form. The molecular wave is different in the two forms, and they can have quite different properties.

All living systems, for instance, are built of left-handed amino acids and right-handed DNA bases. Right-handed amino acids and left-handed bases are non-nutritious and often poisonous.

Just Carbon

The four coupling waves of carbon can link up in chains, and rings of covalent bonds are preeminently suited for system building. This is observed in the two common allotropes of pure carbon, two quite different system waves embracing a huge number of carbon atoms. These are diamond and graphite.

In graphite, the carbon nuclei lie at the corners of 2-D hexagons; in diamond, they lie at the corners of 3-D tetrahedrons.

In diamond crystal, the molecular wave confines the vast number of carbon nuclei into one vast tetrahedral molecule. This is what gives carbon its strength. The electrons are rigidly constrained, so diamond is an insulator. The electrons are in such a stable state that they retard, but do not absorb, photons. Diamond is transparent and slows light down to 43% of lightspeed.

The structure of diamond is a direct reflection of the tetrahedral sp3 orbitals.The graphite structure does not reflect this tetrahedral structure because it involves carbon's penchant for blending waves into new waves.

Delocalized waves

Carbon atoms can use 3 of their 4 valence waves to create 3 'regular' covalent bonds with other atoms. The fourth wave can blend to form a 'delocalized' orbital in which the probability density is spread out all around the molecule.

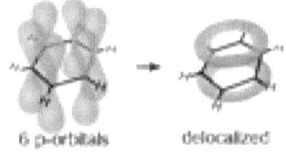

6 p-orbitals   delocalized

The 'aromatic' refers to the alternation of double bonds which allows a single electron wave to span the whole molecule. The simplest example of this is the benzene molecule of six carbons and six hydrogens. Each carbon contributes one of its waves to the delocalized waves, the electrons being delocalized around the whole ring. Compounds in which this delocalization occurs are called 'aromatic' (as they often have strong odors).

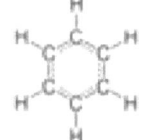

Graphite is a vast flat plane of hexagonal carbons—benzenes without the hydrogens—the whole being one great 2-D molecular wave. The planes stack up but hardly interact, making graphite an excellent lubricant and conductor of electricity. Electrons flow easily along the planes and readily absorb any incident photons, making graphite jet black.

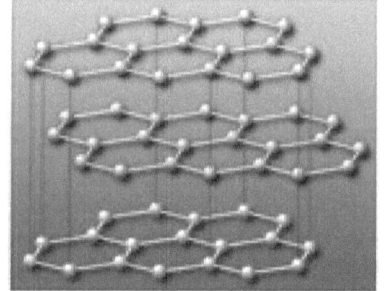

It is the different forms of the wave governing the valence electron density that give diamond and graphite their very different properties, not the subsystems (which are identical in both systems).

Diamond	Graphite
It is the hardest natural substance known	It is soft and greasy to touch
It has high relative density (about 3.5)	Its relative density is 2.3
It is transparent and has high refractive index (2.45)	It is black in colour and opaque
It is non-conductor of heat and electricity	Graphite is a good conductor of heat and electricity

Silicon

Below carbon in the periodic table is silicon. It also lacks four electrons to enter a noble-gas stability, but it is less reactive than carbon and prefers to bond with other atoms rather than with another silicon. Unlike carbon, it is not a subsystem of life but its internal character, particularly its affinity for creating diamond-like structures with oxygen, is essential in the formation of the planets that are the eden and home for life's flourishing. The tetrahedral form in crystal silica is identical to that of diamond except that there is an oxygen bond between each silicon atom. Glass is similar except that the bonds are random and not regular.

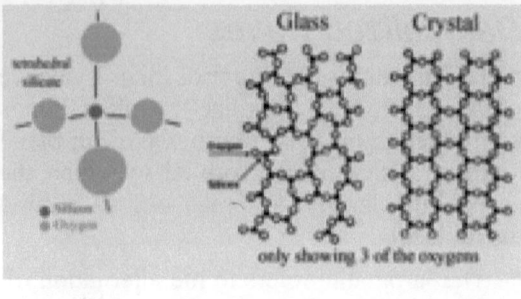

"Silicon is the eighth most common element in the universe by mass, but very rarely occurs as the pure free element in nature. It is most widely distributed in dusts, sands, planetoids, and planets as various forms of silicon dioxide (silica) or silicates. Over 90% of the Earth's crust is composed of silicate minerals, making silicon the second most abundant element in the earth's crust (about 28% by mass) after oxygen."12

IRON

In most elements, the magnetic virtual photons play a small role in the macroscopic properties. It is all too simple for two electrons to rotate so that their magnetic dipoles are aligned, causing attraction, confining all the magnetic virtual photons.

This cancellation is not always possible. In a cold bar of iron, each atom has 1 electron whose magnetic dipole is not aligned and cancelled, and tiny domains in which all these dipoles are aligned. This is state of least energy in which the magnetic virtual photons all add together as boson and create a small magnetic field.

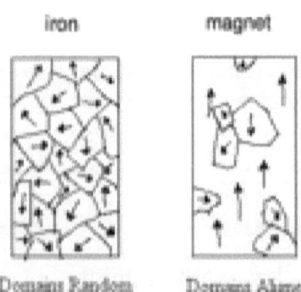

Domains Random   Domains Aligned

(The arrow points towards the north pole)

In an external magnetic field, the domains align with each other. The magnetic virtual photon bosons pile up and reach far into space. The bar of iron has become a magnet.

The probability wave that determines the density of these virtual photons is a "chord" composed of millions of tiny

---

12 http://en.wikipedia.org/wiki/Silicon

generators working together.

In musical terms, we have an array of wave generators creating a 'wave in full.'

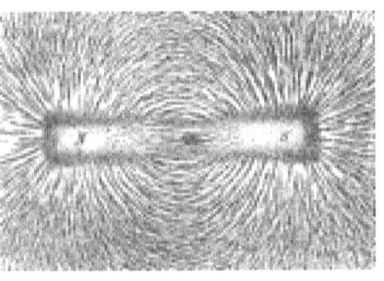

This wave can be visualized by sprinkling iron filings on a paper over the magnets, each speck of iron becoming a magnet and aligning with all the others.

The constructive and destructive flow of 'magnetic lines of force' causes:

1. Like poles to repel

2. Opposite poles to attract

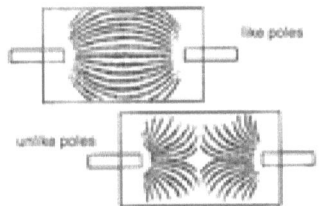

The quantum probability wave and the consequent density of the virtual photons— the magnetic fields—can be experienced directly with two strong magnets by attempting to:

Force two N poles together.

Keep an N and an S pole apart

Electricity

Electricity is the flow of conduction electrons in a metal. The simplest, and first to be noticed historically, is static electricity and its everyday manifestation as static cling and shocks. This is caused by charge separation on a macro scale. When electrons get separated from the nuclei they normally neutralize—such as by being mechanically rubbed off a surface—there is a separation of positive and negative charges. This separation takes energy, and this potential energy creates what is called a 'potential difference' between the two that is measured in volts. If allowed to, the electrons will flow to bring this situation back to overall neutrality, and this flow is called an 'electric current,' measured in amperes (amps), that flows from the negative to the positive.

It is almost impossible to observe static electricity with metals, as the conduction electrons rapidly adjust things back to equilibrium. It is only with materials that do not conduct electricity that this static separation of charge can be maintained.

A van de Graaf generator scrapes off electrons from a source and carries them onto a belt into a sphere, where they are scraped off. The electrons build up until the voltage is so great that the air molecules are ionized and conduct the current as brief spark.

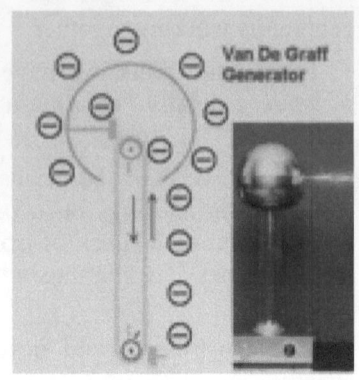

The ionized nature of a plasma makes it an excellent conductor, as do the ions in a melted salt or in solution.

Dynamic electricity

For historical reasons, the flow of 'electric current' is designated as flowing from positive to negative, exactly opposite the flow of electrons. This makes little difference, in practice, as a flow of positive charge in one direction is equivalent to a flow of negative charge in the opposite direction.

In static electricity, the flow of current quickly equilibrates the charge difference, the voltage drops to zero, and the static electricity is 'discharged,' as in the spark from a van de Graaf generator.

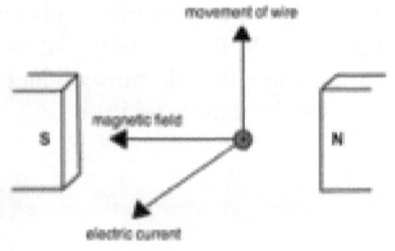

If there is a way of maintaing the voltage, then the current flows continuously, as in the case of everyday electricity. There are two basic ways of generating such a constant voltage: a moving magnet can be used to sweep the electrons along, as an electric generator; or chemical energy can be expended to maintain the potential difference, as in a battery.

When a metal moves past a magnet, the changing magnetic field drags at the electron, and an electric current is generated. The magnetic field, the movement of the wire, and the generated electric current are all at right angles to each other. This is the principle of generators where mechanical motion is converted into electric current.

This generates an alternating current, AC, with the voltage and current in phase and changing as a traveling sine wave.

Complementing this, a flow of electrons through a coil of wire generates a magnetic field. This is an electromagnet. A flow of charge in the earth's outer fluid core generates the earth's magnetic field. The relation of current to field is the 'right-hand rule.'

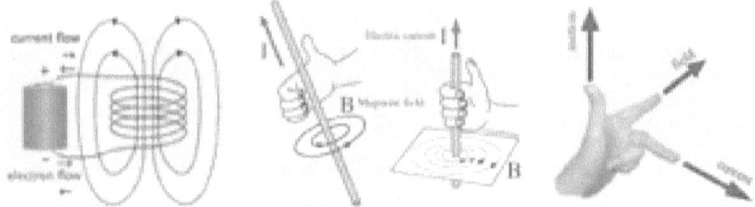

Finally, a wire carrying a current in a magnetic field experiences a force and will move if free to do. This is the principle behind the electric motor. The current, field and motion of the wire are connected by the left-hand rule.

Chemical potential energy can also be converted into electrical potential energy in a battery. Two metals are involved, one with a greater tendency to lose its electrons (more electro-positive) than the other (less electro-positive), such as zinc and copper.

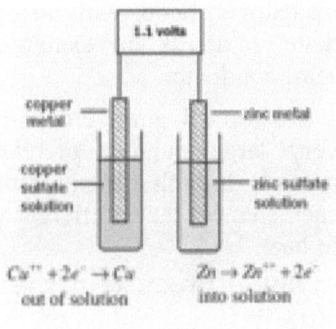

When immersed in a solution of zinc sulphate, the zinc has a tendency to dissolve as its atoms give up their valence electrons and enters solution as zinc ions. Copper, in copper sulphate, has the opposite tendency, it gains electrons and comes out of solution. This disparity shows up as a voltage difference between the two electrodes.

If the two solutions are connected (by a saline bridge or a porous container, for instance), this potential difference can drive electrons around an external circuit. This is a direct current, DC, where the potential difference and the resultant current are constant and in one direction, and can continue until all the zinc has dissolved or all the copper sulphate is converted into zinc sulphate, and the battery goes dead.

Resistance to current

DC electricity is relatively simple, and only one factor, other than voltage and current, needs to be taken into account.

A wire of a pure element has a resistance to the flow of electrons. They bump into atoms and are slowed down, their lost energy of motion adding to the random thermal motion of the atoms. The wire gets hot and its resistance increases.

For a given DC electric potential difference, or voltage, V—set up by a battery or a generator—the amount of current, I, is inversely proportional to the resistance, R, in the external circuit.   This relation is known as Ohm's Law:

$V = IR$

The energy used to overcome the resistance is turned into heat and, in watts, this is:

$W = VI = I2R$

So doubling the current quadruples the heat generated. The chart gives examples of the wide range of bulk resistance to electron flow by pure elements.

Silver and iron have a very small resistance; they are examples of conductors. Sulfur has an enormous resistance and is an example of an insulator. Silicon, with an intermediate value, is an example of a semi-conductor.

To force 1 ampere of current (a very large number of electrons/ second) through a standard block of

Element	Symbol	Resistivity
Silver	Ag	2×10^{-8}
Iron	Fe	1×10^{-7}
Silicon	Si	6×10^{2}
Sulfur	S	1×10^{15}

each element takes different voltages and generates very different amounts of heat.

Silver: $10{-8}$ V & $10{-8}$ W   Silicon: 600 V & 600 W Sulfur: 1015 V & 1015 W

It goes without saying that, with DC electricity, a break in the wire—a gap with a very high resistance—in the external circuit will stop the flow of electrons. This is not the case in AC electricity where the frequency of the sine wave that is the voltage and current is important. The sine wave is most simply described by complex numbers (where, to avoid confusion with the 'I' that is used for current, they use 'j' instead of 'i' to denote the rotation operator).

Along with resistance, there are two other factors in AC circuits that determine the relation between voltage and current—capacitance and induction.

Consider a DC circuit with a break in it. When a voltage is applied, no current will flow. If two metal plates are placed on opposite sides of the gap, however, a current will briefly flow as the electrons build up on one plate until there are so many of them that they repel any more from arriving. Current flows briefly in the circuit (in the opposite direction to the electrons) but nothing crosses the gap.

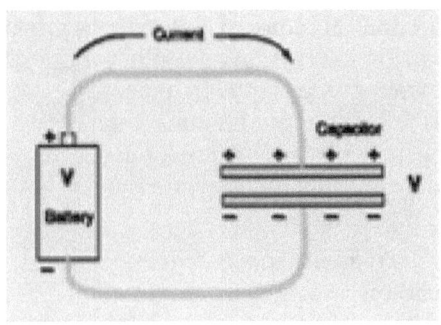

In an AC circuit, the electrons flow back and forth onto the plates with their capacity to hold electrons, and the faster the voltage changes direction, the more electrons can get on before the limit is reached. This is a 'capacitor,' and the larger the plates, and the closer they are together, the greater is the capacity, measured in farads.

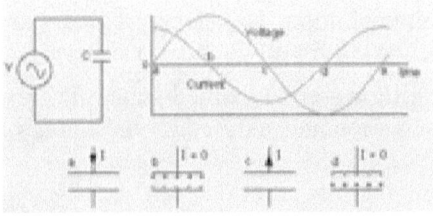

The effect of a capacitor in an AC circuit puts the voltage and current out of phase with each other.

The opposite effect is created by an 'inductor' in the AC circuit, such as a coil of wire. The current through the coil generates a magnetic field, and this change induces a voltage that opposes the flow of current. The inductance, measured in Henries, increases with the frequency and the size of the coil.

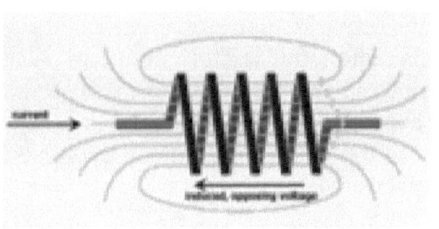

An inductor in an AC circuit, like a capacitor, puts the current and voltage out of phase, but does so in the opposite direction.

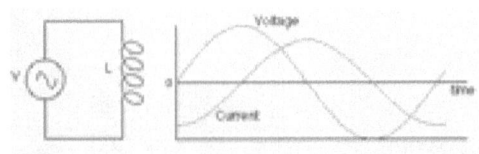

An AC circuit where the phase effects of capacitance and inductance cancel out is said to be a resonant circuit and, as capacitance and inductance vary with frequency, there is a 'sweet spot' frequency at which this occurs called the 'resonant frequency' of the circuit.

EXCITED ATOMS

That concludes our brief overview of the elements and some of their interactions by coupling with electrons and their virtual photons. Atoms are also able to absorb and emit real photons. We have already discussed how thermal photons of low energy move atoms around bodily and alter their kinetic energy. At the other extreme are high frequency UV and X-ray photons that strip electrons from atoms they encounter—they are ionizing radiation.

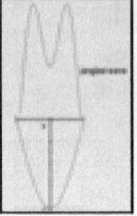 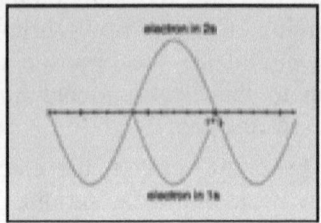

At intermediate values, a photon wave can combine with an atomic wave and put the atom in an excited state, as distinct from the ground state. This can only happen if the photon fits, and is just the right size, for instance, to convert a 1s wave into a 2s wave with an extra internal node. The electrons jitters about in the whole wave, seemingly teleporting from one lobe to another without ever being at the node.

Being in an excited state, the wave quickly reverts to the ground state and spontaneously emits the photon at exactly the same frequency as before. As the smashing together of thermal motion can also kick the wave into such a state, the result is that atoms emit a characteristic set of frequencies at a hot temperature (emission spectrum), and absorb the exact same set of frequencies at a cooler temperature (absorption spectrum).

Each element has its own characteristic 'spectrum,' and this is the basis of spectroscopy that can identify the elemental composition of a lab sample or the composition of the distant stars and galaxies, by the frequencies of light they absorb and emit.

Red Shift

The period of a red photon is approximately twice as long as that of a blue photon. They both have just one quantum action. The blue photon is just more tightly wound. The blue photon has a higher frequency, and a shorter wavelength.

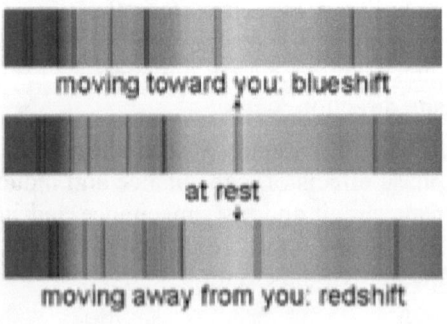

The differences between the photons, however, can be abolished by the relative motion of the source, Sc, and the observer, Ob.

If Sc sends a red photon towards Ob, and the distance between them is constant, Ob can count the arrival of each crest, and calculate the frequency of the photon. It is observed to be red.

If Ob is moving towards the source, however, he will count more crests arriving each second, and see a yellow photon. If the distance is decreasing at an even faster rate, the rate of arrival will be even greater and Ob will now call it a blue photon. This change in observed frequency with relative motion is called the Doppler shift and; as the frequency shifts towards the higher, blue end, this is the blue shift of photons observed when the distance is decreasing; they have a relative velocity towards each other.

The opposite occurs if the distance between is increasing. Less crests are counted per second and the photon is red-shifted. This terminology is used for radio waves and gamma rays, even though the colors are now in the wrong direction. A red-shifted infrared photon actually moves away from the red spot of the spectrum and becomes a radio wave, while a blue-shifted X-ray moves away from the blue spot and becomes a gamma ray.

Combined with the absorption spectrum of the elements which shift against a continuous spectrum, this effect allows a calculation of the relative velocities of stars and galaxies relative to the earth.

This shifting of colors is only significant at high speeds; a substantial fraction of lightspeed is required to make a blue sweater look red.

Expanding universe

Exactly the same effect occurs if the space between source and detector is expanding. The inflation and expansion of the universe is not in space, but in the creation of space. The universe of 10 minutes after the Big Bang was densely filled with gamma rays just after the final annihilation of antimatter —electrons and positrons. There with 100 billion photons for each electron that escaped destruction. These photons are still filling the universe around us, but the expansion of space makes them now look like short radio waves (microwaves), the cosmic microwave background that fills all of space evenly. Nothing has touched these gamma rays since they were emitted, but to us they look like low frequency microwaves.

No energy was lost, of course, they still have their one quantum of action, but it is now spread out over a greater amount of space. A gamma photon has a short wavelength, and it passes by quickly. A microwave photon with a long wavelength takes a lot longer to pass by, as it lazily waves along. In either case, the wavefront is always observed

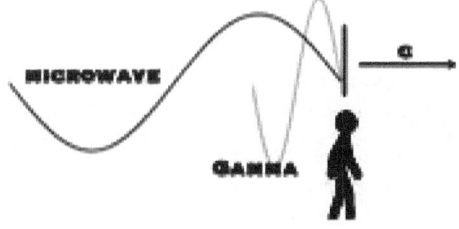

to be moving at exactly lightspeed irrespective of the speed of the observer.

It was an epochal discovery that all the galaxies in the visible universe (except for a few, gravitationally-bound in the Local Group) were all fleeing away from our galaxy, and the further away they were, the faster they were fleeing. They all had a red shift. Speculation that our galaxy was diseased in some way was squelched when astronomers realized that the universe was not, as had heretofore been assumed, eternal and static. They realized that the universe was expanding, it had a beginning and it had a dynamic history.

Rydberg atom

The ground state of an atom is of a very small size. This is relative, of course, for if an electron particle is a grain of sand, and the atomic nucleus Manhattan Island size, the atom would be the size of the earth. But on our scale, they are small enough at between 60 and 600 trillionths of a meter in diameter. The radius of an atom is more than 10,000 times its nucleus, at 2–20 billionth billionths of a meter, but less than 1/1000 of the wavelength of visible light at 400–700 billionths of a meter.

The radius of the orbit can be huge on the atomic scale, the n = 137 state of hydrogen has a diameter ~1 millionth of a meter, a billion times larger, and a billion billion billion times the volume of the ground state atom.

Because the binding energy of a Rydberg electron is proportional to 1/r and hence falls off like $1/n2$, the energy level spacing falls off like $1/n3$ leading to ever more closely spaced levels converging on the ionization energy. These closely spaced Rydberg states form what is commonly referred to as the Rydberg series as shown in the diagram.

This vast distance over which the system wave of a Rydberg atom governs the subsystem—about the size of a bacterium—is an example of the importance of excited states in making it easier to interact with loosely held subsystems.

Light and matter

The wave-confined electron density in atoms can be influenced by photons. The twists of an electron can combine with the twists of a real photon. What happens depends on the frequency of the photon.

An X-ray or gamma ray photon, with a large energy and small time cycle, will rip the electron away and send it flying away at high speed. The now-less energetic X-ray photon continues on its deflected way. The atom is ionized, becoming an ion.

H → H+ + e–

An infrared photon, on the other hand, with low energy and a long time cycle, will just jiggle the whole atom back and forth, adding to its energy of motion as an atom thus adding to its "thermal motion" and increasing its temperate. This is why infrared electromagnetic radiation is also called 'heat rays.

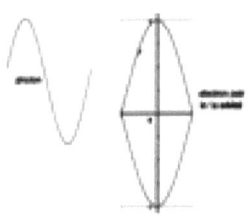

Photons of light, with energy and time cycles in the range of the electron waves in atoms, readily combine with the electron wave. The photon waves and the electron waves momentarily jangle together.

We have already discussed one possibility—the wave combination fits as a standing wave of higher energy. This leads to the specific absorption and emission spectrum of an atom (or molecule)

The other possibility is that the wave does not fit, and the photon wave untangles itself and continues on its way. The photon is retarded in its progress, and this shows up in the everyday world as the refractive index of transparent materials.

Refractive index

The electron-photon wave is not a good fit so the photon continues on its way with exactly the same energy and time cycle as it started out with. This holds for molecular as well as atomic orbitals. For a brief moment, the photon wave was detained in its passage through space.

In passing through the trillions of atoms in a gas, liquid or transparent solid, this retardation builds up and light appears to move more slowly than it would through a vacuum. The ratio of the speed of light in a vacuum to the speed of light in a transparent medium is called its 'refractive index.'

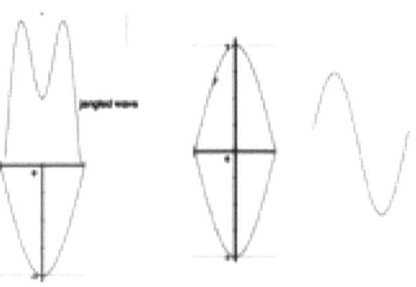

In a gas, this effect is so slight that the index can be taken as 1. The effect can be considerable, however, and mount up to a significant slowing, as in these examples:

Light photons move through water at 75% of light speed, through glass at 56% of lightspeed while the photons are so retarded by the molecular orbitals of diamond that they travel at only 41% of lightspeed.

A charged particle, such an an electron, can be accelerated to almost lightspeed and then sent through a diamond. The electron is now traveling faster than the photons which have, so to speak, no chance to get out of its way. A shock wave of photons builds up and radiates from the path of the electron. This 'Cherenkov radiation' is often how the detectors used in high-energy physics function. A similar shock wave

Medium	Refractive index	Speed of wave
Vacuum	1	c
Air	1.000277	0.999723 c
Water	1.333	0.75 c
Glass	1.8	0.56 c
Diamond	2.42	0.41 c

is created when an airplane or the tip of a bull whip travel faster that the speed of sound, ~760 mph, which is called Mach 1.

A host of photons traveling through space together is called a light ray. The study of these is called 'optics.'

This is one of the earliest 'hard' disciplines to be developed in science. A scientific discipline is 'hard' when it has a well-developed mathematical description; it is a 'soft' science when its description is in a natural language, such as

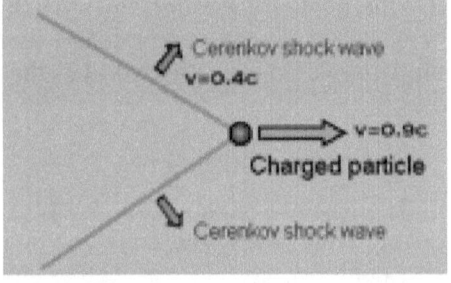

English, with its fuzziness that promotes much hand-waving, or worse.

English has a plethora of words to describe the complexity of things, digital objects. It has very few words to describe the complexity of waves. As wave concepts predominate in a dualistic science, there is a mismatch of vocabularies. Most of the more sophisticated wave words are to be found in music, which is why we introduced the topic so early in the discussion.

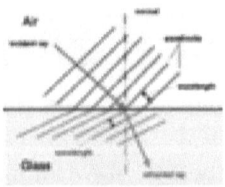

The wavefront of a ray of light passing from air into glass is bent in its course by the speed reduction. This phenomena is summed up in the Law of Refraction:

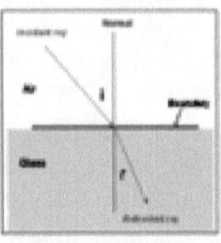

$$\sin i = n \sin r \text{ (where n is the refractive index of the glass)}$$

If the surface by which the ray exits the glass is parallel to the entry surface, exactly the reverse happens, so looking through a pane of glass does not change anything. If the exit surface is not parallel to the entrance, the bending does not cancel out, and we have a lens.

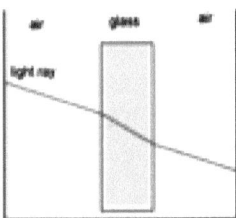

Going from a less dense to a more dense medium, the light is bent towards the normal. Going in the opposite direction, the ray is bent away from the normal. A limit is reached at the incident

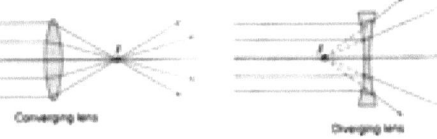

angle that would bend the refracted ray at 90°, and we get 'total internal reflection' where, instead of leaving the denser medium, it reflects off the surface. Diamond, having such a high refractive index and a 'critical angle' at which this occurs, bounces light around inside the crystal before it is released as the sparkle so admired in diamond jewelry.

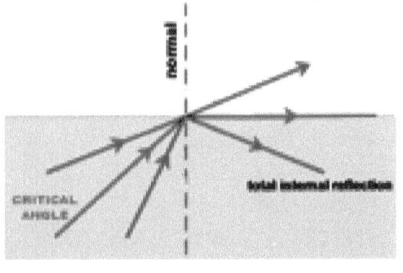

A hot body such as the sun (6,000°) emits a white light, a mix of photons with a spread of energies. As far as visible light is concerned, these photons can be separated into a spectrum of colors by a prism, where the sides are not parallel and the refraction does not cancel out as it does when passing through a pane of glass,

The slowing down of the passage of photons by a transparent substance depends on the energy of the photons, the high-energy (blue) photons being retarded to a greater extent than the low-energy (red) photons. The critical angle for each color light is also different, which explains why a diamond can sparkle in many colors.

5.
EDEN OF
THE MOLECULES

We have discussed how the Logos directs interaction towards a series of edens in which systems can interact and become subsystems of a higher system of a higher 'sophistication.' The higher system has a set of emergent properties inherited from the Logos that were not to be found in the lower systems.

The eden for the creation of the atomic nuclei of hydrogen and helium was the first three minutes of the universe's history.

The eden for the creation of atoms was the 'recombination' of electrons and nuclei about one million years on.

The eden for the creation of the other elements was the aging of two generations of stars over the next eight billion years, and the enrichment of the interstellar gas.

Molecules do form in these enriched clouds, but the atoms are so few and far between that the probability of them getting close enough to each other to interact is on a par with the weak interaction. Many simple molecules, such as water, have been detected by their emission and absorption spectra. Until recently, the rates of reactions in interstellar clouds were expected to be very slow, with minimal products being produced due to the low temperature and density of the clouds. However, organic molecules were observed in the spectra that we would not have expected to find under these conditions, such as formaldehyde, methanol, and vinyl alcohol.

The eden for molecules more complex than these, however, is to be found on planets that concentrate the elements and allow for chemical interactions in abundance.

Before we leave the realm of the universal and focus on our home, we will address a few philosophical questions.

THE ANTHROPIC MULTIVERSE

The philosophical problem is that the natural laws at work in the universe are finely tuned to allow for living systems to emerge and flourish.

The Problem

There are innumerable examples, so we will just mention a few of the 'fine-tuning' situations we have already encountered.

A tiny change in the relative strengths of the four fundamental forces does not allow for the sequence of edens ending in the familiar elements, nor would it allow stars such as our sun to form and burn sedately.

The 'triple-coincidence' that allows for helium to burn into carbon but not all of it into oxygen is plainly exemplary. That massive stars end in a supernova to spread their abundance depends on the precise balance between all four of the fundamental interactions—even neutrinos have a role to play.

The internal characters of hydrogen and oxygen allow for the formation of liquid water which is to life as air is to music. Liquid water in itself has dozens of characteristics that are essential for life.

The equal and opposite H-bonding tendency of nitrogen and oxygen allows for the digital processing of analog information by nucleic acids, and allows for long-term storage and transmission of digital memory. It is the different reduction states, determined by their system wave, that give the closely-related nitrogen and phosphorous their essential, and quite different, roles in living systems.

The facility with which the internal character of carbon allows it to form low-energy covalent bonds with itself and a host of other elements. Most of chemistry (and all of biology) is 'organic,' dealing with carbon-containing molecules; all the rest is 'inorganic chemistry.'

The particular characteristics of left-handed amino acids and right-handed sugars that in combination have emergent properties that are exactly suited to be the basic subsystems of life. No scientist has yet to construct even the simplest metabolism out of D-amino acids or L-sugars, and probably never will as the combination does not have the right character.

Solutions

The problem only gets worse the more we understand about living systems in detail. One current example is the quite unexpected sophistication in the emergent properties of RNA. They can do just about anything because they are chemically versatile. The macromolecule has a finely-tuned set of internal characters that make it the active core of living systems. DNA, in comparison, does few things well as the passive memory core of living systems.

Every year it seems a new type of RNA is discovered, and there are now dozens of known types, each with a different role to play as a unique subsystem of life. They routinely read, write and manipulate digital information about analog forms stored on DNA, are master tweakers of metabolism, and are even considered to be the first macromolecules that interacted as subsystems of a sophisticated system that had the emergent properties we associate with life, such as digital manipulation of memory of analog wave-generators that resonate and structure water.

So, the philosophical question that science has to deal with is: How come the universe is so finely tuned to allow for our everyday existence?

The Abstract realm

We have taken a commonsense approach based on the understanding of the physical world we have already outlined (in reverse):

A system is a set of interacting subsystems coupling with their subsystems.

The external form of a system is the composite probability density of the subsystems.

The subsystems are confined by an internal character, the composite resonance of all the internal waves contributed by the subsystems. This external form of the system, by the Law of Large Numbers over time, is a reflection of the form to the internal character.

The form of the internal character is directly and precisely determined by natural law. As this is more akin to the 'idea of a symphony' than to a simple linear relationship, we place the usual natural laws at the foundation of a very sophisticated abstract entity, we call the Logos.

The commonsense view is that the emergent properties are expressed because they were put there in the elaboration of the Logos. We often use music as an illustration since it has a sophisticated approach to complex waves. When we appreciate qualities such as pathos, excitement and sorrow in a piece of music, we are aware that these were placed there by the craft of the composer. The performance has inherited these qualities from the composer.

Using this as a guide, we expect that the qualities we appreciate in everyday objects—such as being substantial, having finely-tuned valence and chemical behavior—are inherited from the Logos because they were put there in its elaboration. I certainly ascribe the wonderful qualities exhibited by the MacBook on which I am writing—it being a few pounds of aluminum, silicon and plastics—to the expertise of those who drew up the blueprints for their assembly into a MacBook.

Simply put, this again led to a scenario of an Abstract Creator, creating an abstract Logos that acted upon Nothing to create the p-metric history we are discussing and the s-metric we have yet to include. The qualities are expressed by systems because they were put into the Logos with the express purpose of leading to life.

Many intellectuals, however, are more uncomfortable with the concept of an Abstract Creator with a Purpose for creating than they aret with the classically-weird description of the substantial world offered up by modern science. They have offered alternative explanations for the increasingly apparent fine tuning of our universe for life.

Anthropic Principle

Simply put, the 'anthropic principle' answer to the question is another question: "How could it be any other way?" After all, the sun is shining and we are here discussing the question, so how could the universe be any other way. If it were not as fine-tuned as is, we would not be here and the question is moot. This is a little like looking at the Mona Lisa and thinking, "That's delightful; but as it is there before me, how could things be otherwise?"

Such thinking does have its uses, however, as it was just this perspective that enabled Hoyle to think the sequence, "There is carbon, so helium must convert into carbon, there has to be a triple-coincidence of nuclei strong resonances for the triple-alpha process to happen." And so there was. Looking at the painting, we could anthropically deduce, "Someone put paint on canvas to express such personality" and "people have cherished the painting for centuries or it would not have survived in such condition," but not much more than that.

Multiverse

The multiverse explanation for the fine-tuning is that our universe is just one of many universes. The emergent properties are not the result of a universal law, but are randomly assigned and it just so happens that, by chance, all the emergent characters are just right for life.

There are, however, a lot of things that could be quite different, so it takes a lot of them that are not right to account for one that is just right. The discomfort with a purposeful Abstract Creator is apparently less than the proposed number of universes in the multi-universe which is about 10500, a number whose ridiculousness can only be appreciated by reading it aloud: "our universe is one of a trillion, trillion [repeat 'trillion' another forty times] trillion, trillion universes in which the laws are just right for living systems like us."

This perspective is akin to having a profound experience of Beethoven's 9th Symphony and then assuming that it is just one selected from a trillion trillion random compositions.

I personally prefer the commonsense explanation, i.e., the qualities are there because they were put there by great expertise.

That is the end of the philosophical detour, and we now return to what is known about the history that occurred around the sun, our third generation star.

THE EARTH

The gas that condensed to form the earth and sun 5 billion years ago had been enriched by a recent supernova that created an abundance of stable nuclei as well as a host of radioactive elements that have all decayed by now except for a few almost-stable with particularly long half-lives, such as uranium and thorium.

A small fraction of the collapsing gas did not fall into the sun. The metals remained in orbit and combined into molecules, grains and higher aggregates, while the hydrogen and helium was blown away by the solar wind. These aggregates condensed to form the earth about 5 billion years ago.

A collision with another such condensation created the Moon (at ½ the distance it is now). The original crust was like that of Venus—all of one piece—but this impact cracked the shell into tectonic plates that are shifting around to this day. The Moon also stabilized the axis of the Earth's rotation at ~23° inclination, instead of it being all over the place as happened to the planet Mars.

While the water molecule is small, it is chemically attracted to minerals and aggregated with them. Along with the water contributed by the comets during the heavy bombardment phase of the earth's formation, the water condensed on the cooling Earth into oceans that were saturated with reduced, soluble iron in the ferrous state. The Moon raised enormous tides which eroded the land and created great beds of porous clay in the cracks in the ocean floor. These faults between the tectonic plates circulated seawater through the crust as black and white smokers perfusing the clay beds. The intense UV light and lightning drove many organic reactions in the dense atmosphere of carbon dioxide, ammonia and nitrogen.

The history of the universe up to this point has evolved under the direction of the Logos, particularly that aspect we earlier described as the Perfect Wave, i.e., many waves combining over time into a wave with unique properties. As the science describing the formation of 'rogue waves' and 'perfect storms' is in its infancy, this period is as yet only imperfectly described.

But one thing is known for certain. The end result of this 10 billion-year history was an eden perfect for the origin of life to occur with relative rapidity over a few hundred million years. As we shall see, the influence of the Moon was crucial to the rapidity of the origin events that led to life. As very few planets can be expected to have such a large moon, we can expect that life was kick-started on the Earth, and has only slowly developed on the moonless majority. It is to be hoped that most of these moonless planets will have at least developed photosynthetic bacteria and an oxygen atmosphere by the time we get to them.

The emergent properties that were inherited from the Logos made the earth a perfect eden for the emergence of life, entailing the transition from systems characterized by chemistry to a new set of systems that were predestined to be the subsystems of Life.

Biological foundations

In the Book Two we will look at the biological sciences, and the study of living systems. Unfortunately, the 2nd scientific revolution has yet to reach them. The biological sciences have yet to include consideration of the causal internal aspect of things as uncovered by the physicists—the biological sciences are all classical and deal solely with external appearances. So, while the biology of our era is proud of its firm foundations in the 'hard' sciences (those amenable to mathematical rigor), the physics in which current biology is rooted is the classical physics of Darwin's day.

"It is most ironic that today's perceived conjunction between physics and biology, so fervidly embraced by biology in the name of unification, so deeply entrenched in a philosophy of naive reductionism, should have come long past the time when the physical hypotheses on which it rests have been abandoned by the physicists." 13

If the truth be told, much of this slow progress is because the physicists are uncomfortable with their irrefutable discovery of the internal aspect and constantly attempt to explain the basics with classical concepts plus a dash of 'weirdness' that is to be ignored as quickly as possible.

For it was with the greatest reluctance that classically trained scientists faced up to the implication that their description of objective reality was horribly inadequate.

The establishment of the current worldview of physics was not based on theoretical speculation. The current view vanquished the old not for theoretical reasons but because that ultimate arbiter of science, experiment, insisted on it.

"The quantum era had arrived but it did not bring an end to controversy. The interpretation of the new quantum kinematics was, and still is, a source of both conceptual discussion and experimental exploration of its consequences in places where it contradicts deep-rooted intuitions of physicists and others, especially for questions of physical reality and causality. So far, all the experimental tests have decided in favor of the quantum kinematics. More than that cannot be said." 14

---

13 Robert Rosen, Life Itself, Columbia University Press, NY (1991), p. 18.

14 Philip Stehle, Order, Chaos, Order: The Transition from Classical to Quantum Physics, Oxford U. Press (1994), p. 307.

6.
HIERARCHICAL
LOGOS

All that we have discussed so far can be summarized in the general terms of systematics.

An entity that holds together long enough to be dignified with a name is a system. (Short-lived entities are considered resonances or excited states of a 'ground state' system. As witnessed by their role in the 'triple coincidence' that allows for carbon atoms, they can play a role in system-building in an appropriate eden. We shall later encounter another such 'triple coincidence' when three types of procaryotes—bacteria-like living systems—came together to form the precursor of the eucaryotes—all other types of living systems, including ourselves.)

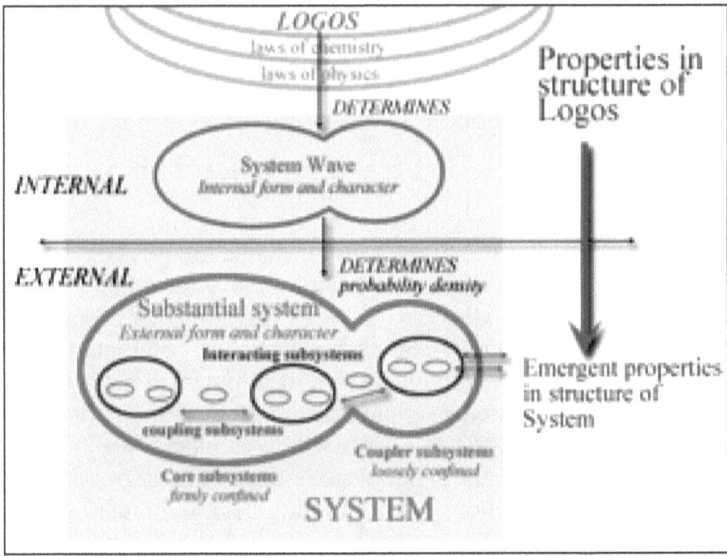

A system has a quantum wave that confines and structures a set of interacting subsystems into the characteristic form of the system. This form gives the system its characteristic properties (which is how we recognize the system). This composite internal hierarchy of subsystem waves is the character of the system, and is what we will be calling the 'mind' of living systems.

The form of this internal character is absolutely determined by natural law and is expressed as the external behavior of the system.

Incomplete confinement of some of the subsystems allows the system to couple with these subsystems, and thus interact with other systems. This is the coupling wave of the system, and it is this that merges with the coupling wave of another to create the interaction wave.

Subsystems, being systems, have a wave that either totally confines its sub-subsystems, the core, or partially confines them, the coupling subsystems.

The waves of the interacting subsystems resonate and harmonize as a wave in full, the mind of the system, and this is given external form as the probability density of the subsystems.

This systematic description of inanimate things creates a hierarchy of systems, with less sophisticated systems being the subsystems of more sophisticated systems. At the base of the hierarchy are the wave and particles of the fundamental entities. At each level a new set of emergent properties are expressed as inherited from the structure of the Logos, and these qualities have been specifically engineered to allow for the higher levels of living systems to emerge, in their turn, as each eden appears in history.

The course of this development involves the emergent properties of molecules in water, so it will be the chemistry of solutions that we will be focussing on. In Book Two we will review the earth's history and the hierarchical sequence of edens in which each step of the system-building occurred.

EMERGENT INHERITED PROPERTIES

We have discussed the origin of the universe, and the origin of the fundamental entities that are the basic subsystems of the physical universe. It took a period of expansion and cooling until the average kinetic energy was sufficiently low enough to allow nucleons to stick together as atomic nuclei (<3 minutes), and for electrons and anti-electrons to complete their almost complete annihilation into gamma rays (<10 minutes). This briefly heated the universe, but it continued to expand and cool until the temperature had fallen sufficiently for electrons to form stable standing waves about the nuclei (~500,000 years). This is the period of recombination, the Origin of Atoms, when the next level of system building beyond fundamental entities began.

Essentially all the nuclei (~90% single protons, ~10% helium-4 nuclei) hooked up with all the electrons and the entire universe underwent a phase transition from a plasma to a regular gas of neutral hydrogen. While a plasma of free charges interacts strongly with all kinds of photons, a gas of neutral atoms is almost transparent photons. The photons 'decoupled' from matter, and have gone their separate ways ever since. This period is also called the dark age as the universe was as transparent as our atmosphere, but there was nothing to see.

Emergent Properties

Something else also happened during the Origin of Atoms. We see a set of emergent properties appearing in the universe, properties that were entirely absent up to that point of history. They emerged on a scene that was designed for their emergence.

We shall discuss a few of the emergent properties (or, more generally, 'emergent qualities') that appeared for the first time in the universe with the Origin of Simple Atoms. (A similar emergence occurred with the Origin of Nuclei, but atoms are more familiar and easier to discuss).

Solidity: Before there were atoms, there was no such thing as a substantial entity in the universe. Hydrogen molecules and helium atoms have the property of claiming a relatively large space for themselves and of excluding other atoms and molecules from sharing that space. The nuclei generator subsystems generate a wave that is a million trillion times larger than themselves, and this is 'fleshed out' as the body of the atom by the probability density of the electrons, the resonator subsystems. Solidity is a quality not observed in electrons and only on a tiny scale by nucleons.

Chemical properties: Electrons and nucleons do not have chemical properties, they do not have the capacity to interact by coupling electrons. (Although much of chemistry and biochemistry talk of hydrogen ion concentration and transport, this is not ever found as a single proton but always as a subsystem of a hydroxonium ion, $H3O+$, or another H-bonding molecule.) All the properties we discussed in our brief survey of the elements emerged during these origin events. (This is a process that continues to this day, so we shall make a distinction between an Origin event (the first of its kind to emerge) and origin events that follow after, a distinction that is irrelevant for non-living systems, but of great significance in living systems.)

Molecular properties: The bulk properties of atoms, such as liquid and solid emerged when the universe was cool enough for aggregations of atoms to stick together. The eden for molecules we are interested in is planets, such as earth, that are in the Goldilocks zone about a star that allows for liquid water.

As we have described, all these emergent properties are a reflection of the external electron density and its confinement by the internal wave; it is the form of the wave that gives these qualities to the atom. The form of all these internal waves—be it system, subsystem or coupling wave—is completely determined by natural law, the lower levels of the abstract Logos.

Hierarchical edens

We have defined the eden for a specific level of system building to be a time and place in which there is an internal coming together as a perfect wave dictated by natural law in which the subsystems are plentiful and the environment is suitable for their interaction as subsystems of a higher system.

This is the cycle, with its levels of increasing sophistication, that unfolds as the Logos directs the course of history and is expressed over time.

Given a population of systems, S1, with a set of interaction capabilities that defines a sophistication of Level One. This population is in an environment, E2, that is just right in terms of conditions such as temperature and concentration for them to have a non-zero probability of interacting and becoming subsystems of a system, S2, with a larger set of more sophisticated interaction capabilities as dictated by natural law.

The S1 systems explore their interactions as their waves overlap and combine in the E2 eden as dictated by natural law.

By the Law of Large Numbers, the non-zero probability of system S2 is expressed, and the first of its kind with its form and properties determined by the Logos emerges in the environment. This is the Origin event for the S2 system.

The S2 system becomes commonplace in the E2 eden and contributes its wave to the development of the perfect wave that is to be the eden for the S3 type of system. For non-living systems, this process of becoming abundant S2 systems is just a repetition of the Origin event of the first, and is just as directly determined by the Logos.

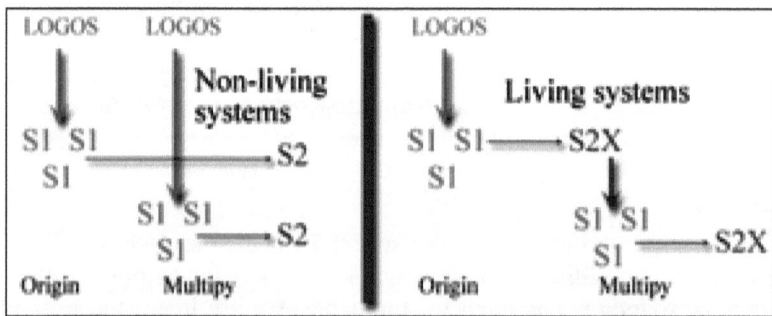

Living systems have a more sophisticated way of becoming abundant. They just copy the first system to make two of them, copy these two to make four of them, and so on in exponential duplication that multiplies S2—one of its emergent properties is the ability to directly assemble system S2 out of the S1 systems without the direct involvement of the Logos. This ability to duplicate is the preeminent emergent property inherited from the Logos that determines the boundary between living and non-living systems. The duplicated systems created by this multiplication are directly determined by the duplication mechanism and only indirectly determined by resonance with the Logos and inheritance of its qualities.

The process then repeats itself. The abundant S2 systems interact and contribute their waves to the perfect wave development of the E3 eden for the next level of sophistication to emerge. Under the direction of the Logos, there is eventually an Origin event of an S3 system, followed by either path of multiplication, followed by Logos-directed interaction and emergence of a E4 eden…

Mutability and stability

There is one more thing to be said about a systematic hierarchy that applies to both the abstract and physical realms: Systems depend on the stability of their subsystems.

In math, this is called consistency. The entire structure of math depends on the lower levels of the hierarchy not entertaining any inconsistency. If, for instance, the lowest levels allow for an integer with an odd and an even count of 2s in its prime factorization, the all of math collapses into a shambles, like Sauron's fortress bereft of its founding ring, and it cannot exist. The existence of the unified systematic hierarchy at the highest levels depends on the immutability of the lower levels.

The same applies to the systematic hierarchy of physical things. A simple example is the dependence on the stability of the carbon atom. Most of the carbon dioxide molecules in the atmosphere have a carbon-12 nucleus as a core subsystem. A tiny fraction, however, have a carbon-14 nucleus (created earlier by the impact of a high-energy cosmic ray proton liberating a neutron which was absorbed by a nitrogen-14 nucleus). This isotope has a half-life of ~5,700 years and will eventually revert back to nitrogen-14. The energy liberated is more than sufficient to disrupt the CO_2.

This hardly makes a difference to the atmosphere, but if the carbon is a subsystem of a large biological molecule, say the DNA that digitally stores a crucial bit of information, the consequences for a complex system such as the human body can be as disruptive and serious as cancer. Our bodies depend absolutely on the stability of the atomic nuclei that are its subsystems, just as math depends on unique factorization.

2ND SCIENTIFIC REVOLUTION

The First Scientific Revolution, started by Newton and ending with Einstein, dealt with the external realm, the only realm in the classical view of the world. This is a world of material particles. Its characteristic is that it is described by real numbers.

The Second Scientific Revolution will complete the picture by adding a description of the internal realm. This complements a world of external particles with a world of internal waves. Its characteristic is that it is described by complex numbers.

Internal wave and external particle—these are the dual characteristics possessed by all things according to modern physics. This duality cannot be described by the familiar real numbers in which all of classical science is expressed. It can only be expressed fully in the complex numbers which have the duality of linear size and circular rotation (or real and imaginary) in their structure and properties.

This new view of the universe is more sophisticated than the old and it adds a whole new level of mathematics that has to be mastered. All the calculation in the new, unified science is done with complex numbers. The result is always a real number, the external aspect that the old science dealt with.

To ask in unified science, "Is an electron a particle or is it a wave?" is akin to asking in mathematics, "Is a complex number real or imaginary?" In either case, the only correct answer is, "It depends."

Since mastery of this new level of mathematical sophistication is not easily obtained (at least for those with a classical education, hence the phrase "quantum weirdness"), it is arbitrarily declared "not necessary" is the less advanced sciences. As scientific thought is hierarchical, with one level of sophistication expressed in terms established by lower levels of sophistication, there is a transition zone from a science that uses complex numbers (which include the real numbers) to a science that uses only real numbers.

This transition zone is currently found, a century after the inception of the modern unified view, in biochemistry which will sometimes be couched in external terms such as 'lock and key' and 'thermal motion' and sometimes in the unified terms of orbital waves and excited quantum states.

classical science external only	current science divided	unified science internal & external
Mind-brain	Mind-brain	Mind-brain
Evolution	Evolution	Evolution
Genetics	Genetics	Genetics
Biology	Biology	Biology
Biochemistry	Biochemistry	Biochemistry
Chemistry	Chemistry	Chemistry
Physics	Physics	Physics

The external foundations of science were laid in the 18th century, while the internal foundations were laid in the 20th. It is important not to confuse the everyday, and fuzzy, use of the words 'internal' and 'external' with the precise definition of these terms used in the sciences. The aspects of nature that are described by complex numbers are 'internal,' while those that are described by real numbers are 'external.'

Logos and Natural law

All the sciences accept the fact that there are abstract laws involved in the functioning of the universe. It is the goal of all the sciences to uncover these natural laws and to derive a mathematical description of them.

The external foundations of classical science required the unspoken assumption that natural laws act directly on the external aspect.

The unified view is more sophisticated. Natural law acts directly only on the internal aspect. For instance, the wavefunction of the electron is exactly determined by natural law and has a precise mathematical description involving complex numbers. This internal wave aspect governs the external probability density of the particle aspect. Within the confines of this external projection of the internal aspect, the particle is found at random.

It is this random aspect that gave quantum mechanics the reputation of uncertainty. This plays only a small role in the systems of everyday life. It is the confinement by the wave that plays by far the main role.

Unified science adds an extra level of causality that is absent in the classical view of the connection between the control by natural law and its expression in the external aspects. As already mentioned, the projection of the internal aspect onto the external is always the absolute square of the internal, so all unified science interest is focused on the internal wave and its developments.

As we are going to show that the laws that govern the internal aspect are hierarchical and not at all like the simple linear image conjured up by the term 'natural law,' we will call the abstract structure of hierarchical laws that emerges in the discussion the Logos.

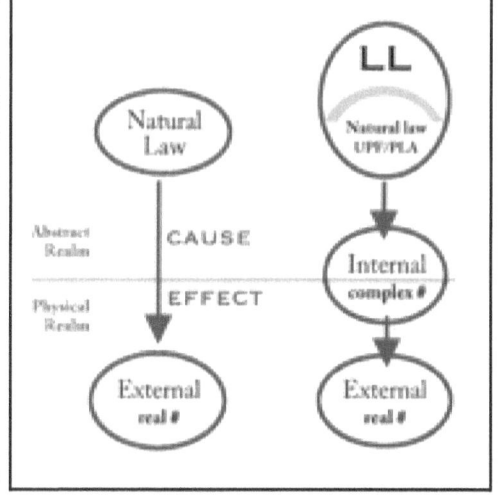

Basic Principles of Unification Science

To summarize, the external interaction of two systems changes the internal waves of both systems. Adding this aspect to the relationship between internal and external, we have now covered the following basic principles of Unification Science:

A system has an internal and an external aspect.

The internal wave is determined by Natural Law.

The external aspect is a set of subsystems.

The internal determines the form of the external over time.

Systems interact externally by coupling with their external subsystems.

Interaction modifies the internal wave, and this change has consequences.

When systems interact in an eden, they can become subsystems of a higher system with   a set of novel properties that are inherited from the Logos This illustration is an outline of the hierarchy of system building by Logos, and a few of the properties inherited from the Logos that will be important in the later discussion. Note that the only difference between atoms and molecules is the number of core wave generators: an atom has only one, a molecule has more than one.

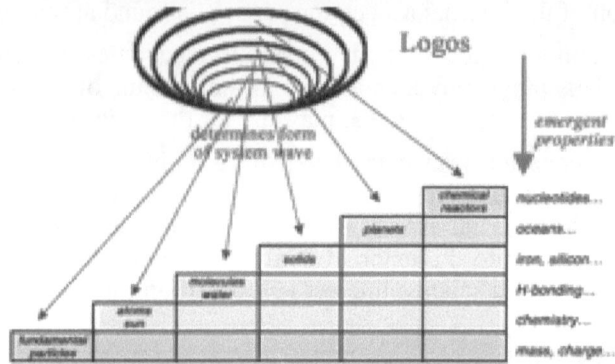

This is a hierarchy of waves with their wave-within-waves structure. It is an intricate structuring of the quantum waves of fermions. The electrons and nuclei are just following along and taking up the density predetermined by the waves.   All the important developments take place on the wave level. It is the quality of the wave that gives properties to arrays of nuclei and electrons.

Unification View

The Logos, like mathematics, inhabits the Abstract Realm. Like the Logos, mathematics is a unified, hierarchical structure, and some of the mathematical structure is intimately tied in with the lower workings of the Logos (and the "unreasonable effectiveness of mathematics in the sciences" is not unreasonable at all).

The view here presented of reality, which includes an abstract Creator and a sophisticated Logos running things, is likely to offend both materialists—who do not admit to an Abstract origin for emergent qualities and only simple laws—as well as Christians and others who consider the Creator to be directly running the Universe. The view is likely to offend both as it implies that both views will have to adjust before we have a unified view in which both are satisfied. To encourage both sides to embrace a larger view we offer each a suitable exemplar, Lord Kelvin for the science-minded and St. Paul, for the religion-minded.

Lord Kelvin was a great scientist of the late 1800s.   Just one of his achievements was to establish the Absolute scale of temperature as commemorated by 'degrees Kelvin' used in all the sciences. His expertise in what was known, however, led him to a famous case of hubris when, in 1899 conference (held just 20 miles from my hometown), he declared that all of science was now known, and all that remained was more and more precise measurements. This was less than a decade before quantum mechanics, relativity and the Big Bang universe entered into the scientific mainstream. A current example of such hubris is Richard Dawkins, who similarly declares that all the basic principles of evolution are understood and all that remains is a more precise understanding of how they worked. We can predict that he will be another famed example of misplaced confidence as the foundation of his view is the Central Dogma of Genetics—that information is read from, but not written to, digital memory. This foundational concept, however, is currently being steadily eroded by the emerging science of epigenetics that does include the writing of information by biological systems to their digital memory store.

The religious exemplar is St. Paul who is so central to Christian theology that excerpts from his letters are regularly read out loud around the world on Sundays. His insistence was not hubris, but a humble admission that "now we see through a glass, darkly" and that there is lot deeper understanding to come in the future.

The ratchet of Interaction

We have already discussed how the wave changes in an interaction. But the wave only determines the density of the valence subsystems, not where they are exactly now. If a hydrogen atom flies past an oxygen, the wave will change but the electron particle might be on the other side. The hydrogen is passed before the electron is on the other side and ripe for coupling.

The wave only determines the probability of a coupling occurring with a valence subsystem. It is the subsystem's situation that determines if the coupling actually does occur.

Coupling is a very black-and-white digital affair: a subsystem was coupled OR a subsystem was not coupled. This is the ratchet of time. It goes as follows.

A system has a structure and coupling wave, A. Time is reversible. All the known laws, and thus the Logos itself, are time reversible.

A system couples with a valence subsystem.

The system wave changes to B. This is the irreversible ratchet of time, A-B. (Going back in A by another interaction is now two clicks of the ratchet, A-B-, not none.)

The system has structure & coupling wave B, time is again reversible.

This is systematic time and it is different from physical time, although the two are often used interchangeably. Physical time, for instance, can go in both plus and minus directions, as it does in matter and antimatter twists.

Systematic time is one-way. Like the integers under addition, it is one way and can only get bigger.

Physical time only began when the particles with rest energy created in the Big Bang had cooled sufficiently so that their kinetic energy fell into the range of their rest energy. Before that, all had been moving at essentially the speed of light in space and there was no extension of physical time.

A cosmic ray proton in the current era is an example. It can have a kinetic energy of over 1020 eV, far greater than its rest energy of 109 eV. The rest mass of the proton is inconsequential and it moves at lightspeed through space and so is motionless in the internal aspect of physical time. When it interacts with an atmospheric nucleus, however, and ejects a neutron before continuing on with a little less energy, it moves forward in external systematic time.

SAMPLING THE WAVE

As we have established, all things are composed of external fundamental particles that are confined by an internal wave. As the fundamental particles are, in all cases, identical, all the qualities of the many disparate things are to be found in the internal wave.

The divide between the inanimate systems described so far, and the living systems that will be dealt with in Book Two is to be found in the internal wave, not the external particles. This necessitates a further extension of the science of sound waves to set the foundation for the discussion to follow.

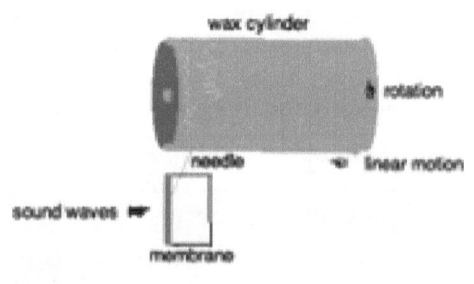

The great American inventor, Edison, was the first to put this aspect of science to practical use when he invented the first method of sound recording. The method involved a thin membrane that moved back a forth with the change in air pressure, and attached to this was a needle that made a scratch on a wax-coated cylinder.

This cylinder was put into two motions. It rotated at a moderate, constant speed, and it also moved slowly from left to right. In quiet air, the needle inscribes a smooth helix about the cylinder. Placed in the room with the tuning forks resonating to the A-major chord, the needle scratches out a wavy line just like our machine did.

Note what Edison has done with his phonograph that combines constant linear motion with constant circular motion: he has taken a sample of the resonance filling the room. This is a huge and complex structure and the phonograph sensor, the membrane, takes only a very small sample of this three-dimensional vastness.

It outputs a simple compound sine wave, a linear wave in just one dimension.

We have sampled the huge complex wave-in-full and output a much simpler, smaller wave. This is called 'down-sampling' a wave. We can call this down-sampled, simpler wave the wave-in-image of the wave-in-full. The wave has been recorded as a scratch in the wax, so the process can be called 'recording' a wave-in-image to linear memory.

Edison did not stop there. He first turned the soft wax impression of the wave-in-image into a hard wax form and then—brilliant when you think about it first—he ran the apparatus backwards.

The back and forth movement of the needle in the groove was transferred to the membrane that vibrated along with the needle.This membrane transferred its vibrations to the air and, voila, a sound could be heard filling the room. The sound in our example would be the A-major chord.

We have turned a small, linear, one-dimensional wave into a huge three-dimensional resonance of the air in the room. The simple wave memory has been transformed into a huge and complex wave that is a replica of the original wave-in-full. The wave-in-image has been up-sampled to the wave-in-full. This is "replaying" a simple, linear wave-in-image into the large complex resonating replica of the original wave-in-full.

The simple mechanism of Edison's phonograph is an example of a reversible 'transducer' that can down-sample a wave-in-full into a wave-in-image record and when run in reverse, replay the recorded wave-in-image and recreate a replica of the wave-in-full.

The transducer connects the wave-in-full form of the extended, multiple resonators to a wave-in-image memory of the form. This memory can be replayed to recreate the wave-in-full form of the resonators.

We will call the mechanism through which the complex wave is down-sampled and, when reversed, up-samples the complex wave again, a simple 'transducer' that can run in either the read or write direction.

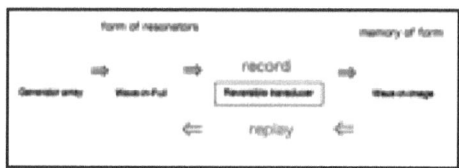

Edison's simple method was quite capable, for the first time in history, of recording talk, music and song, and playing it back again as many times as required.

More sophisticated versions of the transducer soon emerged where the writing and reading of the linear wave became separated, specialized and sensitive. The culmination of this was the black vinyl record that dominated the 1950s through 80s, with a 'single' disc that rotated at 45 rpm and the much larger album, or LP, that rotated at 33 rpm. Here, the wave movement of the needle was electrically amplified before reaching the membrane or loudspeaker.

A symphony

Let us go back to the array of sound generators. We discussed a very simple example: three tuning forks together generating an A-major chord. Now we can discuss the array of generators that allow us to create the most sophisticated waves-in-full: the symphony orchestra.

For our example we will use the climactic finale of Beethoven's 9th Symphony, the sublime Ode to Joy, the anthem of the European Union.

For specificity, we will locate the performance in Carnegie Hall, a vast hall with wonderful acoustics. The array of sound generators includes:

400 human singers in banks of sopranos, altos, tenors and basses. 400 string instruments in banks of 1st and 2nd violins, violas, cellos and double-basses. 300 woodwinds in banks of bassoons,   oboes and clarinets, 300 brasses in banks of trumpets, trombones, French and English horns, flutes and piccolos. Drums, pianos, a couple of harps.

Each one of these sound generators, even when sounding the same note, resonates with a different mixture of standing waves: the 1st harmonic, the 2nd, 3rd, 4th, etc. It is this different mix of harmonics that gives an instrument its 'timbre', its unique sound that makes an A on the piano sound so different from an A on a violin. It is what makes one voice sound different from all others.

At the climax of the symphony, pretty much every generator is going full out, pouring out its unique blend of harmonic sound waves. It is quite overwhelming.

The mass of air in the hall is resonating with a constantly changing, extremely complicated, wave within wave within wave. This is the wave-in-full.

There are also an array of samplers, the audience. Each tiny eardrum is picking up a down-sample of this wave-in-full. It transmits a wave-in-image through a series of tiny bones to the inner ear. This does a simple Fourier Analysis of the wave-in-image and sends the results to the brain where the wave can be recreated as we 'hear' the sound. The ear is an example of a sophisticated one-way transducer that can only record but not replay the wave.

Edison's example of a simple reversible transducer is quite capable of recording the wave-in-image of the symphony, and of playing it back, although at a much lower volume in its simplest form.

Memory of a perfect wave

A perfect wave-in-full can be recorded as a wave-in-image. If the process is reversible, and if there is memory, the perfect wave can be recreated. We shall continue this discussion when we introduce the difference between nonliving systems which get their wavefunction from the Logos, and living systems. Living systems get their wavefunction from the Logos in the origin event, and then store a wave-in-image in memory which can recreate the perfect wave and also be duplicated.

Analog and digital

So far, all the waves we have discussed have been continuous, and they smoothly change their form and shape. They are examples of what we will call 'analog form.' The resonance of the generators, the wave-in-full structuring the form of the resonators, and the memory of it as a wave-in-image are all examples of analog form.

The final thing we need to know about music does not involve analog form. Rather, it involves 'digital information' that is not continuous, i.e., it comes in distinct bits called 'notes' in music and 'quanta' in physics.

Linear digital information always involves a reading convention that is followed by a set of wave generators, which together structure a mass of resonators into a characteristic wave-in-full form.

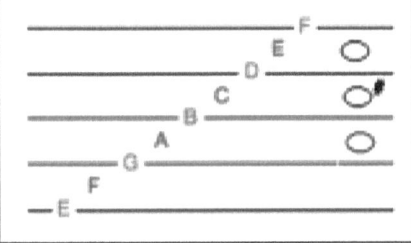

For sound waves, this convention is called 'music notation.' It specifies the frequency of the 1st harmonic to be sounded (along with any characteristic higher harmonics) and its duration.

Each named note has its own

place on a 5-lined stave, either in a space or on a line. A simple oval instructs the generators to sound an A, a C# and an E together for a whole beat, the basic rhythm of the piece.

For a complex piece, like Ode to Joy, each bank of musical generators gets its own line of digital instructions, one page covering a few bars of music. Each bank reads the score at the same rate, and when instructed to sound, does so. This is parallel processing of digital information—there are many parallel lines each being read by an array of different wave generators.

The constructive interference of all these waves is the singular wave-in-full we call a symphony performance.

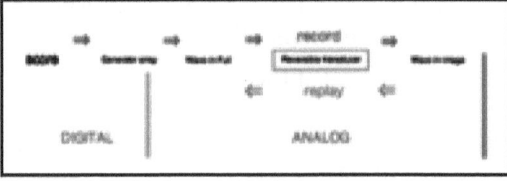

For all its magnificent complexity, the Edison phonograph is quite capable of down-sampling the performance into a simple, 1-D wave-in-image.

This is parallel digital information being read in time by an array of banks of generators. The recording and replaying of a down-sampled wave-in-image by a transducer is as before.

This was the situation of sound recording until the 1990s with the LP record (except for the specialization of the transducers into separate, irreversible recording and replaying mechanisms).

The final step was taken with the advent of the CD and the storage of information about the wave-in-image in a digital form.

The wave-in-image from the transducer is fed into an electronic device that measures the height of the wave many thousands of times a second. The analog-to-digital device outputs this as a stream of digital numbers that are written to the CD. An example covering just a fraction of a second would be:

99 98 99 96 90 80 70 71 72 74 75 78 60 50 40 23 etc.

This string of digital numbers is eventually written to a compact disk, a CD, in a binary code, utilizing only the digits 1 and 0.

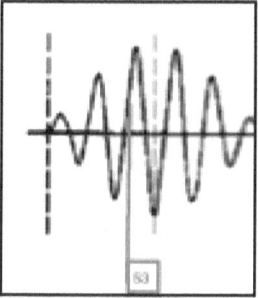

The CD player reverses this process, it is a DC-AC transformer. Each number tells it how much electricity to send to the amplifier and loudspeaker. As the thousands of numbers per second pour into the player off the CD, the current to the loudspeakers is a recreation of the wave-in-image. The loudspeaker turns this into a wave-in-full that recreates the symphony.

It is easy to see that the process can be made into a circle if the same reading convention is used for the digital information at either end—the instruction of the generators and the recreation of the wave-in-image. An example of this is the MIDI system for music recording and playback that uses the same coding convention for both.

Waves and computers

Using the term 'computer' in its broadest sense, we have four different types of computers involved in the creation of music, and the recording and replaying of music. Using DC to stand for digital information, and AC to stand for analog wave forms, we have:

DC/AC computers. The array of primary sound wave generators (dozens of musical instruments, singers, etc) are of this type. Using a universal reading convention, they take a multitrack digital input (a score, a MIDI track) and convert it into an analog wave that gives form to a mass of resonators. The CD player is also a DC/AC computer: it turns the stream of digital numbers into a voltage wave, the wave-in-image, which drives the secondary sound generator, the loudspeaker. It has been found that a better recreation of the wave-in-full is obtained by taking two samples of the wave offset from each other. These two wave-in-image recordings are used to energize two loudspeakers separately, resulting in stereo sound. Dolby sound recording involves at least four channels and four loudspeakers. In general, however, there are many more primary generators than there are secondary generators.

AC computers. The simple, reversible phonograph transducer is one of these. In the record direction, it takes a down-sample of the wave-in-full form of a mass of resonators, and outputs a simple wave-in-image. In the replay direction, it inputs the wave-in-image and outputs the wave-in-full that, relatively faintly, recreates the symphony. The phonograph is a ±AC computer. In later, more sophisticated developments, the transducers became irreversible. The writing direction now involves a membrane in a microphone jiggling magnets past each other. The tiny currents this sets up are electronically amplified into an electric current whose alternations are the wave-in-image. It is this current that drives the writing needle in the soft wax or serves as input to the CD writer. The microphone/amplifier combo is a + − AC computer. The reading step involves a needle vibrating in the groove attached to magnets whose current changes, yielding the wave-in-image, are amplified and used to drive a loudspeaker that recreates the wave-in-full. The output wave-in-image from the CD player is treated similarly. The amplifier/loudspeaker combo that up-samples the wave is a − + AC computer.

AC/DC computers. The CD writer is an example of these. It takes a wave-in-image and samples the size of the wave repeatedly. It outputs these almost instantaneous sizes of the wave as a linear string of measurements written in a standard code. A MIDI recorder does something more complex, but its final output is also a sting of binary bits, just using a more sophisticated code.

DC computers. These have digital input and digital output. A computer that can take a MIDI-encoded file and print out a multi-part score for an orchestra is an example of this. This is a feat well with the capabilities of the DC computer I am writing this on, a MacBook Pro, if it were provided with the appropriate software program. This would be a digital program that is fluent in two coding conventions, both MIDI and musical notation, taking the MIDI input and translating into musical notation. All current computers are of the DC variety.

It might appear that my Mac is also capable of outputting AC in the form of sound, photos and movies. This is but an illusion: the output is purely digital and discontinuous. It is the resolution that gives the impression of AC, not DC. It is the thousands of discontinuous pixels on the screen—each either red, green or blue—that "fools" my eyes into seeing smooth and continuous colors. It is easy to see such pixels with a magnifying glass.

Music and Matter

This concludes our discussion of the basics of music and its recording.

We have discussed constant rotary and linear motion, and the sine and cosine waves, both traveling and standing. We have discussed how sound is described as combinations of sine waves, and how it is generated by resonating mechanisms. We have discussed the connection between a digital score, an array of sound generators and the wave-in-full that structures the air in the concert hall.

Further, we have discussed the recording and replaying of the wave-in-image as well as four types of computers that variously manipulate AC forms and DC information. This is summarized in the diagram.

The ratchet of systematic time is sometimes called the 'collapse of the wavefunction.' This is not a very good name because it implies that the wavefunction, the term quantum scientists use, collapses and disappears. This does not happen. Rather, the wave changes from one wave to a another, usually from an extended wave to a localized wave.

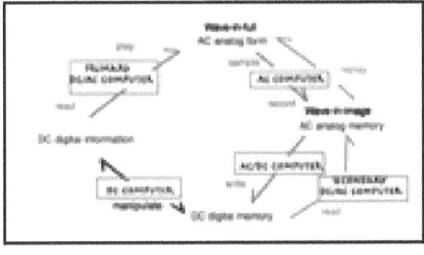

The term arose from the earliest experiment that detected the wave aspect of matter, the slit experiment. This involved a simple set-up.

A particle (such as an electron) is shot out of a gun in a vacuum, passes through a barrier with two slits in it, and hits one of the detectors arrayed on the other side.

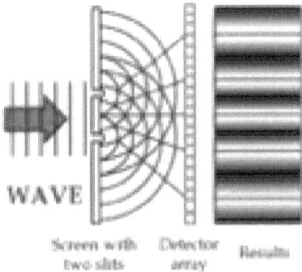

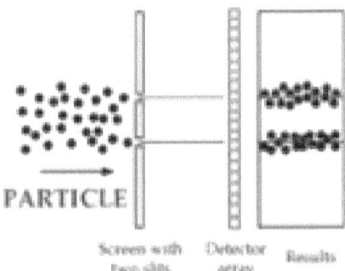

WAVE

Screen with two slits   Detector array   Results

PARTICLE

Screen with two slits   Detector array   Results

The electron behaved as a wave would, not as was expected from a particle. This was hard to understand for those who considered an electron a solid particle, and that 'solid' was not fundamental quality, but one inherited from the Logos by composite systems.

This wave-like behavior remained when only a single particle traversed the apparatus at a time. Each electron ended up in a single detector, say a crystal of silver iodide in a photographic plate. The wave-pattern emerged when thousands of electrons were sent one after another. The electron wave was interfering with itself.

Each electron wave, after being split in two by the slit, continued on its interfering way until it reached the detector array. At this point the wave is spread out over the centimeters of the entire array. Hitting one of the detectors, this extended wave immediately disappeared and the electron appeared in the local detector. This is where the term 'collapse of the wavefunction' originated, to describe the disappearance of the extended wave.

It is not really a disappearance, however, so the name is unsatisfactory. What actually happened is that the electron wave began to merge with all the waves of the silver atoms, coupling with virtual photons occurred with an atom where the particle happened to be in its jittering, and the electron wave snapped to being one with the atomic wave of the silver atom in the detector crystal.

The single electron wave passed through the slit into a double-lobed electron wave with a sizable node. The two lobes interfered with each other into a wave pattern of nodes and antinodes covering the detector array. Whilst at a location in the wave, a coupling of virtual photons with the ion of silver at that location occurred. This alteration of subsystems snapped the extended, rippled electron wave into a subsystem wave of the silver atom, a black speck in the silver iodide crystal. At all times, a single electron particle is jittering about in a single electron wave that determines its density over time. The change in the wave-that-fits from plate size to atomic size is instantaneous, but there is always a wave at all times.

This type of demonstration of the wave aspect has recently been reported for buckminsterfullerene, a spherical molecule of 60 carbons. The size of this decidedly a bit of 'matter' makes it difficult to digest when it 'travels through both slits at the same time.' This behavior is difficult to 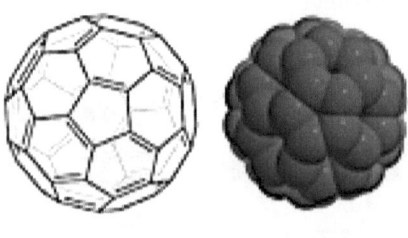 explain by those who think the property of being solid is a given, and not a sophisticated quality inherited by sophisticated system waves from the Logos. The demonstration has yet to be performed on something as ponderous as a tRNA, but a tRNA is more wave than particles.

www.ingramcontent.com/pod-product-compliance
Lightning Source LLC
Chambersburg PA
CBHW032000170526
45157CB00002B/487

* 9 781300 731450 *